T0177277

Environmental Footprints and Eco-design of Products and Processes

Series Editor

Subramanian Senthilkannan Muthu, Head of Sustainability - SgT Group and API, Hong Kong, Kowloon, Hong Kong

Indexed by Scopus

This series aims to broadly cover all the aspects related to environmental assessment of products, development of environmental and ecological indicators and eco-design of various products and processes. Below are the areas fall under the aims and scope of this series, but not limited to: Environmental Life Cycle Assessment; Social Life Cycle Assessment; Organizational and Product Carbon Footprints; Ecological, Energy and Water Footprints; Life cycle costing; Environmental and sustainable indicators; Environmental impact assessment methods and tools; Eco-design (sustainable design) aspects and tools; Biodegradation studies; Recycling; Solid waste management; Environmental and social audits; Green Purchasing and tools; Product environmental footprints; Environmental management standards and regulations; Eco-labels; Green Claims and green washing; Assessment of sustainability aspects.

More information about this series at http://www.springer.com/series/13340

Subramanian Senthilkannan Muthu
Editor

Assessment of Ecological Footprints

 Springer

Editor
Subramanian Senthilkannan Muthu
Head of Sustainability
SgT Group and API
Kowloon, Hong Kong

ISSN 2345-7651 ISSN 2345-766X (electronic)
Environmental Footprints and Eco-design of Products and Processes
ISBN 978-981-16-0098-2 ISBN 978-981-16-0096-8 (eBook)
https://doi.org/10.1007/978-981-16-0096-8

© The Editor(s) (if applicable) and The Author(s), under exclusive license to Springer Nature
Singapore Pte Ltd. 2021
This work is subject to copyright. All rights are solely and exclusively licensed by the Publisher, whether
the whole or part of the material is concerned, specifically the rights of translation, reprinting, reuse
of illustrations, recitation, broadcasting, reproduction on microfilms or in any other physical way, and
transmission or information storage and retrieval, electronic adaptation, computer software, or by similar
or dissimilar methodology now known or hereafter developed.
The use of general descriptive names, registered names, trademarks, service marks, etc. in this publication
does not imply, even in the absence of a specific statement, that such names are exempt from the relevant
protective laws and regulations and therefore free for general use.
The publisher, the authors and the editors are safe to assume that the advice and information in this book
are believed to be true and accurate at the date of publication. Neither the publisher nor the authors or
the editors give a warranty, expressed or implied, with respect to the material contained herein or for any
errors or omissions that may have been made. The publisher remains neutral with regard to jurisdictional
claims in published maps and institutional affiliations.

This Springer imprint is published by the registered company Springer Nature Singapore Pte Ltd.
The registered company address is: 152 Beach Road, #21-01/04 Gateway East, Singapore 189721,
Singapore

This book is dedicated to:

The lotus feet of my beloved Lord Pazhaniandavar

My beloved late Father

My beloved Mother

My beloved Wife Karpagam and Daughters—Anu and Karthika

My beloved Brother—Raghavan

Everyone working with various industrial sectors to make our planet earth SUSTAINABLE

Contents

About the Editor

Dr. Subramanian Senthilkannan Muthu currently works for SgT Group as Head of Sustainability and is based out of Hong Kong. He earned his Ph.D. from The Hong Kong Polytechnic University and is a renowned expert in the areas of Environmental Sustainability in Textiles and Clothing Supply Chain, Product Life Cycle Assessment (LCA), Ecological Footprint and Product Carbon Footprint Assessment (PCF) in various industrial sectors. He has five years of industrial experience in textile manufacturing, research and development and textile testing and over a decade of experience in life cycle assessment (LCA), carbon and ecological footprints assessment of various consumer products. He has published more than 100 research publications, written numerous book chapters and authored/edited over 95 books in the areas of carbon footprint, recycling, environmental assessment and environmental sustainability.

Ecological Footprint of the Life Cycle of Buildings

Cristina Rivero-Camacho, Juan Jesús Martín-Del-Río, Jaime Solís-Guzmán, and Madelyn Marrero

Abstract In recent years, construction and use of buildings in Spain accounted for more than 40% of energy consumption, 36% of CO_2 emissions, 50% of material extractions from the earth's crust and 25% of the solid waste generated. Optimizing the resource requirement in existing buildings is a key step in the task of reaching the community targets for 2030: 27% improvement in energy efficiency and 40% of reduction of greenhouse gases. For this, it is necessary to analyse them through environmental and economic indicators, so that the magnitude of the impacts can be qualified and quantified from the Building Life Cycle (BLC), from its conception, through the extraction of raw materials, for the manufacture of the materials, until the demolition of the building. The paper focuses on the study of residential buildings in Andalusia, covering its entire useful life, from its conception until the time of its demolition or end of life. As a case study, the Ecological Footprint (EF) indicator is applied to a residential project, to calculate the impacts during the BLC.

Keywords Ecological footprint · Emissions · Construction · Building life cycle · Resources consumption

1 Introduction

Construction is responsible for the consumption of more than 40% of natural resources and 30% of energy and produces more than 30% of greenhouse gas emissions, and is responsible for a significant part of wood and water consumption in the world [1]. The reason for the considerable impact of construction is to be found in the building processes, from the manufacture of materials, through their construction and subsequent use, and ending with demolition. In view of the need to implement improvements in the environmental aspect of construction, it is necessary to have verifiable and reliable indicators that are sensitive to changes.

C. Rivero-Camacho · J. J. Martín-Del-Río · J. Solís-Guzmán (✉) · M. Marrero
Departamento de Construcciones Arquitectónicas II. ETS Ingeniería de Edificación, Universidad de Sevilla, Avenida Reina Mercedes n°4, 41012 Sevilla, Spain
e-mail: jaimesolis@us.es

© The Author(s), under exclusive license to Springer Nature Singapore Pte Ltd. 2021
S. S. Muthu (ed.), *Assessment of Ecological Footprints*,
Environmental Footprints and Eco-design of Products and Processes,
https://doi.org/10.1007/978-981-16-0096-8_1

On December 2, 2015, the European Commission approved the so-called Circular Economy Package. This document sets out guidelines to ensure sustainable growth by using resources and waste in a more intelligent and environmentally friendly way. The main idea lies in the search for sustainability in the construction sector, optimizing the use of construction products, planning in such a way as to minimize the production of construction and demolition waste (CDW), the consumption of resources and even providing, where appropriate, for modular construction, the use of industrialized construction elements, possible deconstruction, and the use of products that can be reused or recycled after their first use.

Therefore, in order to improve the environmental performance of buildings, it is necessary to analyse them by means of environmental and economic indicators, so that the magnitude of the impacts can be qualified and quantified throughout the building's life cycle (BLC), from the extraction of raw materials, for the manufacture of construction products, to the demolition of the building. However, the possible combinations in building designs are high, together with the large number of cases within the duration of the BLC, making it difficult to analyse.

Currently, there is a tendency to use simple methodologies, as society can more easily understand them. Among these, the ecological footprint (EF), carbon footprint (CF) and water footprint (WF) are the most prominent [2]. This success is due, first, to the fact that the results they produce are understandable by non-scientific society [3], and second, due to its ease of application in environmental policy and decision-making [4]. An easy-to-communicate and reliable indicator can influence consumer decisions, legislation and regulation [5], allowing the assessment and comparison of human demand for resources and the production of raw materials, as well as the capacity to absorb the carbon generated in production processes [6].

Therefore, the indicator EF is selected for application to building projects. The indicator EF was introduced by [7], who measured the EF of humanity and compared it to the carrying capacity of the planet. The EF is defined as the amount of land that would be needed to provide the resources (cereals, feed, fuelwood, fish and urban soil) and absorb the emissions (CO_2) from humanity. Methodologies that include several indicators may be preferable because they avoid overlapping impact categories [8]. EF can be studied by category (classifications of different productive territories, see symbolically in Fig. 1), which helps in the identification of the main sources of impact [9]. The methodology currently applied to calculate the EF [10] is set by an international body called the Global Footprint Network, which brings together researchers and sustainability experts from around the world [11]. On the other hand, the EF presents weaknesses such as the aggregation of factors from various sources into a single indicator, normally only giving an overview of all impacts within an activity or productive sector [4], and aggregation is subjective, based on the assumptions in order to express all results in a single unit [12].

This work is part of the research developed in the ARDITEC group. The model previously developed by the authors, among others, in Spain [14] will be taken, thus trying to calculate the footprint of any project from the design phase. The research group has been working on calculation models for the different stages of the BLC, urbanization [15], construction [2, 16], use and maintenance [17], and rehabilitation

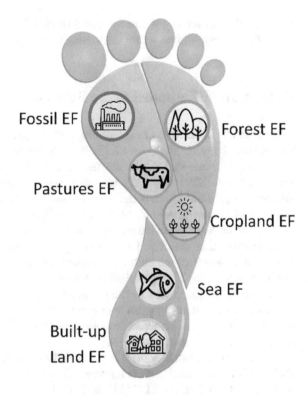

Fossil EF

Forest EF

Pastures EF

Cropland EF

Sea EF

Built-up
Land EF

or demolition [18], developing these in an independent but coordinated manner. Finally, Freire et al. [19] developed a methodology capable of determining the EF of the elements that form part of the traditional construction cost databases, approaching it from a new "environmental budget" perspective, using the tools normally employed in building project budgets.

This research responds to the complete evaluation of the BLC through the application of the environmental indicator EF. It is proposed with a methodology based on building project quantities, to which environmental coefficients calculated of the different categories will be applied. With this idea of environmental budget (where materials, manpower and machinery are considered), we want to promote good management and study in the design phase of the building project, to opt for solutions "cheaper" environmentally and that, consequently, reduce the environmental impact in the BLC.

For the development of the research, a bibliographic review of the available models and environmental indicators applied to the different stages of the BLC has been previously carried out, both in an isolated way and as a whole. With all this, the objectives and methodology to be followed in order to respond to this research work have been established. Once the limits of the system are established, the duration of the phases is applied to an actual project of a residential building in Andalusia (Spain). The results obtained are classified and analysed according to various criteria in order to extract the impact points and the conclusions of the research work carried out according to the objectives set for calculating total impact of BLC with EF indicator.

2 LCA and EF Applied to the Building

To measure the interaction of the building with the environment and to identify the load balance between all stages of a building's service period, life cycle analyses (LCA) are recommended. LCAs are regulated by the international standards ISO 14040 [20] and ISO 14044 [21]. By taking into account all the flows exchanged between the product/system under analysis and the environment, LCAs provide an overview of the environmental performance of the object under study and help to support circularity between the different product systems. It has been widely applied in the construction sector and is increasingly used as an advocate for decision-making at all levels of the built environment, material [22, 23], systems [24, 25], entire buildings [26, 27] and neighbourhoods [28, 29]. It is true that the application of LCA methodologies to the construction sector is complicated and varies according to several studies, this is due to the fact that existing standards have not been able to establish an exact methodology, and therefore researchers, use their own interpretations of these standards [30].

A large number of studies strive to estimate the environmental impact of buildings [31] have been presented in several reviews focusing on LCA [32], life cycle energy analysis [33] or CF of the BLC [34], or a combination of these [35, 36]. There are numerous analyses of the environmental impact of buildings, although only a few cover the entire BLC (see Fig. 2), and the vast majority are limited to the analysis of energy consumption. Most of these building LCA studies can be found in various reviews conducted in recent years [30, 37, 38].

Several researchers have chosen to use the EF indicator in their studies, having applied it to high-rise districts in Tehran [39], farmers' houses [40], hotels [41] and the renovation of a hundred-year-old house [42], in addition to having developed

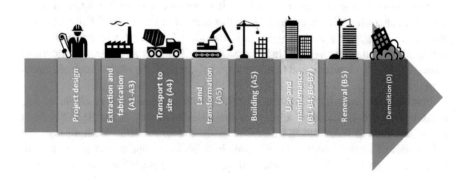

Fig. 2 General diagram of the Building Life Cycle, BLC (self-made)

a tool for estimating EF and CF of buildings [43]. Finally, [44], analysed the BLC (project, materialization, use and demolition) and its study according to EF (energy, resources, CO_2 and solid waste), applying it to an exhibition centre in Wuhan (China). Bastianoni et al. [45] calculated the EF of two Italian buildings, taking into account the embedded energy of the materials and the construction process (estimated as 5% of the total energy of the materials). The results are reflected in land for CO_2 absorption, forest land (for wooden materials) and the area occupied by buildings.

Despite the fact that the results of these studies often remain barely comparable due to the myriad of assumptions and decisions that must be made in the evaluation process (life of the building and materials, maintenance operations, energy consumption of the building, building typology, calculation formulas, etc.), the following findings are generally identified:

- The manufacturing and construction phase of the building's life cycle, concentrated in a short period of time (1–2 years), causes the most intense environmental impact, mainly due to the consumption of concrete and steel for the structure, which represents a high percentage of the emissions produced during this phase [46, 47]. This impact is diluted, the longer the life of the building is considered, however, decisions made during this phase greatly influence the results for the remaining phases of the building's life cycle.
- The use and maintenance phase is generally responsible for 80–90% of the CO_2 emissions generated during the building's life cycle [48], almost 60% of which is caused by energy demand for heating and air conditioning [49]. Its duration, more than 50 years, makes the reduction of emissions in the operational phase the main objective to be pursued.

The reduction of energy consumption during the use and maintenance phase should be achieved through decisions made during the design phase, which generally involve the use of materials with higher built-in energy. This means that, in nearly zero energy buildings, emissions during the construction phase represent a higher percentage of the total emissions during the whole BLC [50]. Therefore, once the operational energy has been reduced, the attention of researchers should focus on the development of new insulation materials that require less energy to manufacture [51]. Although the stages of land use transformation and recovery or demolition produce a smaller environmental impact in comparison, it is interesting to note their influence on the biocapacity of the plot, as well as their relationship with other stages of the BCL.

With these findings in hand, one cannot ignore the importance of anticipating the impacts that will occur as a result of the design of the building. The knowledge generated by the significant number of studies on environmental impact is subsequently reflected in calculation tools, generally oriented towards the environmental certification of buildings as a means of communicating their results to society. The development of models to measure the sustainability of buildings has been encouraged in recent years to promote policies that require minimum sustainable behaviour for buildings.

In this sense, this research studies each of the stages of the BLC, from design to demolition, the end of life phase of the building, in order to respond to the environmental assessment of the entire BLC.

3 Objective and Methodology

Taking into account the initial planning, the objective of this work is to develop a model that allows the evaluation and quantification from the design stage of the building, the material and energy resources present in each of its life stages, so that it is possible to predict what would be its environmental impact in its entire BLC.

All of this is planned from a budget structure that allows for a double evaluation, economic and environmental. This identifies the location of the impact points, as well as facilitating their understanding and application by the technicians of the sector as it is a structure with which they feel familiar and which also allows their transfer to other documents and agents involved in the building projects.

Since the decisions taken in the design stage will have a direct impact on the rest of the building stages, it is intended that the method proposed will evaluate through the EF indicator the environmental impact of the project under consideration in a quick, efficient and practical way, with the greatest possible reliability of results. In addition, the model developed could enable the use of the BLC evaluation as an assistant in the design of projects. These objectives will be carried out according to the methodological diagram developed in Fig. 3.

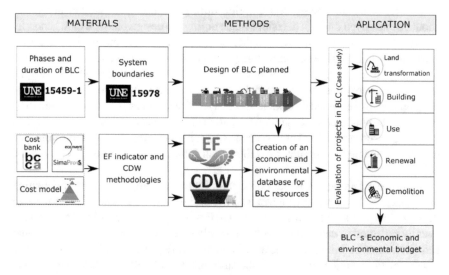

Fig. 3 Methodological diagram for BLC evaluation. CDW stands for construction and demolition waste (self-made)

3.1 *Materials*

The methodological map starts in the materials section, aimed at collecting all the data and preliminary considerations necessary for a good approach to the BLC to be applied in the model. For this purpose, the following subdivisions will be taken into account.

(a) *BLC stages and duration*

In this section, the time limits of the system are defined by determining the moment when the BLC begins and ends; that is, the scope is delimited in a longitudinal direction. Likewise, the limits between the phases of the BLC will be marked, which should establish a clear and strict separation, so that there is no duplication in the quantification of the different impacts.

In this sense, the limits are clear in the case of new buildings, whose cycle begins with the transformation of the land for construction and ends with the completion of the demolition work. The starting point of the use and maintenance phase is established after the completion of the construction work and the commissioning of the building. This, being available to be occupied, use stage is the longest of its life cycle, which is usually estimated between 40 and 100 years, and ends when the building's useful life.

Defining the moment when its useful life ends is not so simple. For example, Adalberth [52] divides the BLC into three phases, subdivided in turn into the manufacture of the materials, their transport to the site, the construction of the building, the occupation of the building with an intermediate period dedicated to the renovation, to finish with the demolition and waste or recycling. This division by phases is quite generalized, with slight variations from one study to another, and limits the use and maintenance phase to energy consumption and renovation or remodelling work.

On the other hand, the CEN/TC 350 "Sustainability of construction works", UNE-EN 15804 [53], recommends the consideration of not three, but four stages in the life cycle of buildings, separating construction from product manufacturing [26], so that there would be a first phase of production, which includes: extraction of raw materials, transport to factory and manufacturing; a second phase that defines construction, transport and construction or installation processes on site; the third phase that focuses on use including, maintenance, repair and replacement of products, reforms, operational energy (heating, air conditioning, ventilation, hot water and lighting) and operational water consumption; and the fourth and last phase aimed at the end of life, deconstruction, transport, recycling or reuse, and disposal.

These models have in common that the life of the buildings ends with their demolition, but what would happen if instead of being demolished it was decided to rebuild? In the event that the renovation route was chosen, the question arises as to whether the post-rehabilitated building could be considered the same building to which its useful life has been extended or, on the contrary, it would be a new building to which its life cycle begins again.

In response to the questions raised, for the present work, it is assumed that the end of life of the building occurs when it ceases to exist, i.e. after its demolition. From the point of view of the proposed renovations, they are considered part of the maintenance of the building, so that the maximum performance of the original building can be obtained, prolonging its useful life in a moderate way. With this in mind, every time the building is renovated, the improvement of its energy efficiency will be taken into account. This will be motivated by the changes, both in their constructive solutions and in their installations, being these modernized with respect to the original ones, with the consequent change in the patterns of consumption.

In order to establish the optimal moment when building renovations will be considered, the standard UNE-15459-1-energy efficiency of buildings is taken as a reference. Procedure for the economic evaluation of the energy systems of buildings BLC assessment method. Inputs and outputs for the calculation of the building's energy efficiency [54]; and the UNE-15686-1-useful life of materials [55]. These must be complemented with the Technical Building Code in Spain [56], which establishes, based on Eurocode EN-1990, that residential buildings must be designed for a duration of 50 years or 100 years if they are for public use. For this reason, the study is set at 100 years, considering an increase in the useful life of the building due to proper maintenance and renovations. Based on these reference standards, it is estimated that the renovations will take place when it is 20, 40 and 70 years old, on a cumulative basis, adapting the items included in each work to the duration of the various installations and components of the building. Since the installations are the elements that require most frequent renewal, they will be intervened in each renewal project. From 40 years old and onwards, the replacement of the woodwork and the relevant repairs to the enclosure are included. The latter will be repaired again 30 years later; in the renovation at 70 years old, where together with the previous elements mentioned, structural repairs will be included, as well as the installations that are more difficult to replace, such as the sewerage system.

With all that is set out in this section, and the considerations of durability of the materials presented in the Larralde research [57], the longitudinal limits of the scope of this study are established, which give rise to the designed BLC, which is detailed in Sect. 4 of the case study. In summary, the BLC begins with the urbanization and construction of the building, which would include all the environmental impacts associated with the extraction, manufacture, transport and installation of the materials. Once the building is occupied, the 100-year use phase begins. This stage will be subdivided into four periods of use due to the different renovation works planned at 20, 40 and 70 years old. Finally, demolition is considered at 100 years old, when it is estimated that the building will no longer be fit for habitation.

(b)   *System boundaries*

Once the time limits have been established, the limits of the BLC are set in terms of the factors to be taken into account in this study as opposed to those that would not be quantified for the environmental assessment. These are its transverse boundaries, which will form the separation between our system and other production sectors, such as the furniture industry, household appliances, construction materials manufacturing and waste treatment plants. This will be essential to avoid double quantifications and overlaps between cycles—building–surrounding relationship.

For the definition of the limits, the ISO 14040 [20] standard on LCAs and the UNE-EN 15978 [58] standard (sustainability of construction works, assessment of the environmental performance of buildings, calculation method) will be taken into account.

The limits of the system are defined for each of the phases of the BLC in three sectors: industry, construction and occupants. The impacts produced during the manufacture of furniture, appliances, decoration and other objects belonging to the building's occupants are assigned to the corresponding industrial sector, not to the building. Similarly, the impacts generated by the occupants, such as food consumption, mobility and municipal solid waste, are their personal impact [59]. With regard to the wastewater and waste generated by other consumables (coal, biomass, etc.), it is considered part of the impact caused by the occupants, not the building itself.

In summary, the limits of the system are the resources consumed by the building, namely the building materials and their transport, the machinery used and the manpower required to carry out the work, as well as the electricity and water consumed during the use stage.

(c)   *Systematic work classification and cost database selection*

The selection of the correct systematic work classification in the building will be fundamental to establish a robust and reliable structure for the incorporation of environmental data into the building budgets and to make possible the unification of criteria for all the stages, that is to say, to create a construction cost database (which includes economic and environmental costs) for the BLC. The model is based on the Work Classification System (WCS) of construction budgets for each stage of the life cycle, urbanization, construction, use and maintenance, renovation and demolition. The automation of data and processes is advanced in Information Technology (IT) that provides great advantages in predictive analysis and in the sector prevails the classification systems of construction information (SCIC) as management tools, in a review by Freire [60], are highlighted among others, MasterFormat [61], Uniformat [62], Standard Method of Measurement of Civil Engineering [63], CI/SfB [64] and the Uniclass [65].

All these bases are proposed as ideal tools for the realization of the economic quantification or budgeting and also as an integrating element since its system of decomposition and hierarchization allows the introduction of a standardized process. All classification systems have the same basic concept, to divide a complex problem into simpler parts that can then be added, without overlap or repetition, to define the complete development of the projects. In Spain, construction cost databases have their own classification and its scope is usually the geographical setting, the Institute of Construction Technology of Catalonia [66], PRECIOCENTRO in Guadalajara [67], BPCM Madrid [68], BDEU in the Basque Country [69], BDC-IVE in Valencia [70] and the ACCD in Andalusia [71, 72].

The model used in the present work is the ACCD [72], widely used for estimating costs in construction by the Andalusian public administration. Based on a hierarchical and tree structure with defined levels, where each group is divided

Fig. 4 Systematic
classification structure with
examples (self-made from
[73])

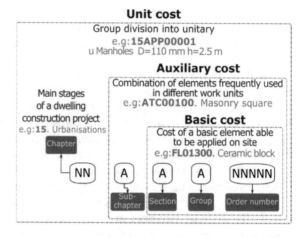

into subgroups with homogeneous characteristics. The divisions are called chapters, and each represents a construction process, such as demolition, earthworks, foundations, water disposal, structure, partitioning, roofing, installation, insulation, finishing, carpentry, glass and polyester, cladding, decoration, development, safety and waste management [72]. The divisions and examples are listed in Fig. 4.

The classification system materializes the codification of each concept, which means that each code corresponds to a concept and each concept to a unique code, allowing precise identification. Among other advantages, it also facilitates IT management and solves the location of concepts in the budget structure [19].

The reference cost database is most widely used in the region, which has been published continuously since 1986. Its structure is created with clearly defined levels, in which from the apex of the hierarchy one descends to the lower levels, dividing each group into subgroups of homogeneous characteristics (see Fig. 5). Thus, the base of the pyramid is formed by the supply costs (SUP), which connect the system directly to the factor in the market, manpower, materials, machinery, subcontractors, etc. At the top of the structure are the contract quantities or direct costs (DC), which connect the economic information with the product markets (residential, offices, schools, housing developments, etc.). The structure is completed by inserting between the extremes, depending on the degree of detail sought, intermediate levels [73], such as basic costs (BC) (distributed mainly according to the three natures mentioned, materials, machinery and manpower), auxiliary costs (AC) formed by the union of BC previously described with the quantities appropriate to their typology and function; unit costs (UC) formed by the union of BC exclusively or in combination with AC. This cost coding hierarchy is shown in Fig. 5. At the top of the pyramid shown in Fig. 5 are exogenous costs such as industrial profit (IP), taxes (VAT) and overheads or general expenses (GE) of the construction company.

All these characteristics facilitate the incorporation of the environmental cost based on the same hypotheses and contours defined in the calculation of the economic cost. Therefore, the ACCD will be the ideal one to use in the model to be developed, as it is considered to be of proven strength and appropriate for the objectives set.

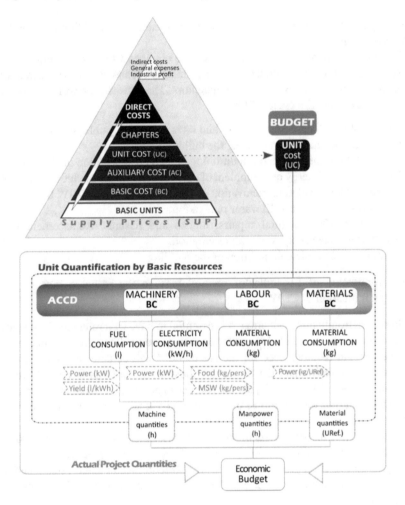

Fig. 5 ACCD structure. Obtaining resources of materials, machinery and manpower from the building budget (self-made from [74])

3.2 Methods

This second level is intended to provide a response to the achievement of a reliable economic and environmental database for application to the BLC of any project. In this respect, the subdivisions detailed below shall be taken into account:

(a) *Design and programmed activities for BLC*

Together with data obtained in the previous divisions, the characteristics and activities or actions programmed for each stage of the BLC are established, obtaining the design of a "standard BLC" with which to evaluate the building projects from the design stage.

For each of the phases, information will be obtained from the execution budget and the plans for the calculation of resources, manpower and machinery that compute in the methodology of environmental indicators. The construction works carried out at each stage of the life cycle will be:

- Urbanization: Road works, sewage and its installations, public services, etc.
- Construction: The construction of the building.
- Renewal 20: Renewal of air conditioning and DHW generation installations.
- Renovation 40: Energy re-equipment of the roof and facades (including windows), including their insulation, renovation of air conditioning and DHW installations. Wet cores, floors, doors. Elevator renovation.
- Renovation 70: Structural repairs, cracks and fissures. Replacement of all installations, electricity, water and sanitation.
- Demolition: Complete demolition of the building.

(b)   Application of EF indicator methodology to the resource inventory and quantification of CDW.

In the environmental evaluation, it is essential to have well referenced the sources used to obtain the associated impacts for each resource. Therefore, in this section, the sources and methods used for the calculations will be defined. In this line, the necessary formulation and the flow chart illustrating the steps to be followed are presented, in addition to specifying the sources to be used for obtaining the required auxiliary data and the definition of the necessary coefficients and transformation factors.

In the present model, international databases of LCAs of construction products [75] and the EPDs available on ECO-Platform [76] (www.eco-platform.org/), a European platform for EPD programmes in the construction sector that is established with the aim of an implementation of EN 15804 [53] with mutual recognition among the members, are used.

Among the different LCA databases, the ecoinvent database [77] was chosen [78], implemented in SimaPro [79] and developed by the Swiss Centre for Life Cycle Inventories, due to its transparency in the development of processes (reports, flowcharts, methodology, etc.), consistency, references and standing out for the fact that it merges data from several databases of the construction industry [75]. From this database, a series of "environmental families" have been obtained which will be responsible for assigning to each basic cost their corresponding impact units according to their similarity.

One of the main objectives to be achieved is that the evaluation model is a standardized and reproducible model in all phases of the BLC, which is achieved thanks to the fact that resources are systematically quantified and classified in the project budget. This has allowed the methodology of the environmental indicators to come closer to the concepts used in the terminology of the work budgets, thus facilitating the understanding of the model by the technicians of the sector and encouraging the application of the same.

The analysis of environmental costs has been carried out by adapting the evaluation model presented by Solís-Guzmán [80], where the EF indicator was applied to the building sector. Figure 6 shows the different concepts evaluated, classified on various levels (impact sources and footprints). This methodology of adaptation of the environmental assessment model for the building sector is the basis of the environmental study developed in this research. This methodology has evolved since its original publication through the work of several researchers [2, 16, 14].

This methodology applies to the resources used (energy, water, manpower, building materials, etc.) and waste generated in the construction of residential buildings, generating what has been called the "Resource Inventory", this being the starting point for environmental assessment. This quantification includes all the entries of the building extracted from the budget, which makes it possible that the model can be applied in all the phases of the BLC, since the established cost database, structured according to the ACCD, allows to quantify the resources used in the actions to be carried out on the building (Fig. 6). In summary, the EF calculation model will determine the total footprint, which in turn is composed of different partial footprints: (fossil) energy, grasses, sea, crops and forests. These come from the impacts generated by resources, energy and the generation of CO_2 emissions. The concepts are catalogued in the scheme as intermediate elements, and their mission is to transform consumption into elements to allow the calculation of the different footprints that make up the global footprint.

The following is an explanation of each of the elements and the approximations adopted for their calculation, following the base methodology for the evaluation of

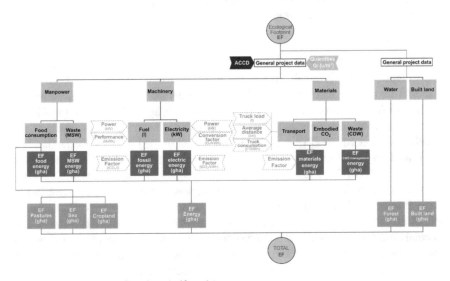

Fig. 6 EF Methodology flowchart (self-made)

the EF [2, 16, 80] and the scheme shown in Fig. 6. The summary of all the equations used for the calculations is shown in Table 1.

Firstly, indirect resources have been defined, which in traditional construction budgets correspond to manpower, materials and machinery. These cause impacts on the work through the energy consumption of the machinery used on site (fuel or electricity), the impact of the workforce (through the quantification of working hours) and the consumption of construction materials (during their manufacture, transport and installation).

1. Manpower

Next, the analysis of the impacts generated by the workers on the site is carried out, generation of Municipal Solid Waste (MSW) and food consumption. The mobility of the operators to the work site or place of work is excluded in order to adjust the methodology to the proposal in the UNE-EN 15978 [58], for the Life Cycle Analysis (LCA) of the building. The EF produced by the workers' food is considered the worker's energy source and is obtained from Eq. 1 in Table 1.

The footprints are generated according to the type of food (meat: EF of pastures; fish: EF of productive sea; cereals: EF of farmland), with the natural productivity and equivalence factor of each productive territory. The crop equivalency factor is 2.51; for pasture, it is 0.46; for forest, it is 1.26; for sea, it is 0.37; and for direct area of occupation, it is 2.51 gha/ha [83]. All foods will also produce an energy footprint, due to the energy consumed in their transformation. The equivalency factors used [11] are implicit in the calculation of EF (EF/person and year) for each country, see Table 2. It has been taken into account that breakfast and lunch (which are made on site) represent approximately 60% of the daily diet of an adult [84]. To obtain the footprint of the MSW, an average generation coefficient per worker and the emission factor of their treatment are used (Eq. 2 in Table 1).

2. Materials

To obtain the EF of each product or building material used in the project, first the weight of these is obtained, and the CO_2 emissions contained in each kg of material are applied according to the LCA databases. Once all the CO_2 emissions of each material have been obtained, Eq. 3 in Table 1 is applied to obtain the EF. Up to this point, the environmental impact of the materials during the life cycle of the cradle to door is covered. To evaluate the A4 aspect of the UNE-EN 15978 standard [58], an analysis of the transport of the material is also carried out, establishing approximations of the distance covered by the means of transport, see Table 3 [19].

The first thing to take into account is the means of transport to be used; in our case, it is by truck, whose capacity is defined in tons and average consumption of diesel, also taking into account the CO_2 emissions for each litre of fuel consumed. The second aspect to be considered is the distance to the work site from the factory and/or the warehouse of each material; taking the following approximations, in Andalusia, it is considered that most of the materials are manufactured in that area, for which the average distance is 250 km. In the specific case of concrete, a maximum distance of 20 km will be considered, according to the considerations of EHE-08, Art. 71,4.

Table 1 EF calculation equations (self-made from [18])

	Equation
Manpower	
EF_{foi}: Partial Ecological Footprint of food consumption in EF category i (gha/yr)	
$EF_{foi} = \frac{H_w}{H_d} \cdot 0.61 \cdot \frac{EF_i}{365}$	1
H_w: Total number of hours worked per year (h/year)	
H_d: Number of hours worked per day (8 h/day)	
0.61: Breakfast and lunch as a percentage of the total daily food intake of a Spanish adult (61%)	
EF_i: Footprint of food consumption in EF category i (gha/person)	
365: Days in a year	
EF_{MSW}: Partial Ecological Footprint of MSW management (gha/yr)	
$EF_{MSW} = H_w \cdot G_w \cdot E_{MSW} \cdot (1 - A_{oc})/A_f \cdot EQF_{ca}$	2
H_w: Total number of hours worked per year (h/yr)	
G_w: Hourly waste generation (0.000077 t/h) (EUROSTAT, 2015)	
E_{MSW}: Emission factor of MSW (0.244 tCO_2/t) [81]	
Construction materials	
EF_{ma}: Partial Ecological Footprint of consumption of materials (gha/yr)	
$EF_{ma} = \sum(C_{mai} \cdot E_{mai}) \cdot (1 - A_{oc})/A_f \cdot EQF_{ca}$	3
C_{mai}: Consumption of material i per year (kg/yr)	
E_{mai}: Emission factor of material i (tCO_2/kg)	
EF_{tr}: Partial Ecological Footprint of the transport of materials (gha/yr)	
$EF_{tr} = \sum(\frac{W_{mai}}{T_{cap}} \cdot D_{ma}) \cdot T_{con} \cdot E_f \cdot (1 - A_{oc})/A_f \cdot EQF_{ca}$	4
W_{mai}: Weight consumption of material i (t/yr)	
T_{cap}: Truck capacity (t)	
D_{ma}: Average distance (km)	
T_{con}: Truck consumption (l/100 km)	
E_f: Emission factor of fuel (tCO_2/l)	
Machinery. EF_{mc}: Partial Ecological Footprint of machinery (gha/yr)	
$EF_{mc} = \sum(H_{mci} \cdot C_{fi} \cdot E_{fi}) \cdot (1 - A_{oc})/A_f \cdot EQF_{ca}$	5
H_{mci}: Hours of use of machinery i (h/yr)	
C_{fi}: Consumption factor of machinery i (l/h or kW)	
E_{fi}: Emission factor of fuel used by machinery i (tCO_2/l or tCO_2/kWh)	
Water consumption. EF_{wa}: Partial Ecological Footprint of water consumption (gha/yr)	
$EF_{wa} = C_{wa} \cdot EI_{wa} \cdot E_{el} \cdot (1 - A_{oc})/A_f \cdot EQF_{ca}$	6
C_{wa}: Water consumption per year (m^3/yr)	
EI_{wa}: Energy intensity of drinking water (0.44 kWh/m^3) [82]	
Built-up land. EF_{bl}: Partial Ecological Footprint of built-up land (gha/yr)	
$EF_{bl} = S \cdot EQF_{bl}$	7
S: Total surface occupied by the building or parcel (wha)	
EQF_{bl}: Equivalence factor of infrastructure land (2.51 gha/wha) [11]	

Table 2 EF of daily food consumption per year and person [84]

Country	Cropland EF (10^{-3} gha)	Pastures EF (10^{-3} gha)	Sea EF (10^{-3} gha)	Fossil EF (10^{-3} gha)
Spain	1.45	0.27	0.41	0.49

Table 3 Data for calculating the impact of transport [19]

	Concrete	Other materials
Truck load capacity (kg)	24.000	2.000
Distance to factory (km)	20	250
Average diesel consumption (litres/100 km)	26	26
Diesel emissions (tCO$_2$/litre)	2,62E−03	2,62E−03

With these data, the tons of CO_2 that would be involved in the transport of each material can be obtained, applying Eq. 4 of Table 1, converting these data into EF—a figure that is passed on directly to each basic cost.

As an example, we can see in Table 4 the most representative families of materials in building projects together with their environmental impacts in EF, associated per unit of reference.

3. *Machinery*

The impact caused by the use of machinery is analysed, specifically by its energy consumption (both fuel and electricity), linking it to the engine power. The consumption of the machinery is calculated, and then, the EF of the fossil fuel consumption is obtained from Eq. 5 (Table 1).

To obtain the fuel consumption, the "Machinery Manual" prepared by SEOPAN [85] is used, where the technical data of different models and typologies of machines on the market are collected. Choosing the most unfavourable consumptions, the classified machinery is analysed, where a coefficient is applied to the power of each

Table 4 EF of the main families of materials (self-made)

Material	EF (hag/t)
Soil	0.005
Wood	−0.483
Concrete	0.057
Asphalt	0.098
Ceramics	0.107
Aggregates and stones	0.005
Metals	0.907
Plastics	0.898
Glass	0.30
Mortars and pastes	0.294

engine to obtain the litres of fuel consumed, differentiating whether the machine consumes diesel or petrol. Once the litres of fuel consumed are known, the coefficient is applied to obtain the amount of CO_2 generated by one litre of fuel [86].

A similar approach is followed for the consumption of electrical machinery used on site, analysing the engine power, and the hours of use, obtaining the total kWh consumed [60]. To these data is applied the coefficient which indicates the CO_2 emissions generated for the production of one kWh of energy by the Spanish electrical system [87], that is, the GHG emissions, or global warming potential (GWP) and transforming them into tons of CO_2 equivalent.

4.   Water consumed on site

To obtain the water consumed in the irrigation of concrete during construction of the building structure, the investigations of [2], obtaining a coefficient that relates the m^3 of water per constructed surface. Finally, the coefficient that indicates the electrical energy necessary (kWh) to obtain one m^3 of water is applied to the total water consumed, as indicated in Eq. 6 (Table 1).

5.   Land consumption

The EF methodology takes into account the land that is directly occupied, as it will be biologically unproductive from the moment it is developed. The indicator defines two possible types of territory, forests or crops. In our case, the plot is considered agricultural. In the present analysis, as in previous works [2, 14], according to Eq. 7 in Table 1.

6.   Construction and Demolition Waste (CDW)

CDWs are the remnants, waste or demolition materials on site. Consequently, they will not have environmental impacts themselves as their original impact was accounted for in the entry of the new materials. Instead, it will be necessary to determine the CDW hours of machinery and manpower required for collection and transport to the landfill or recycling plant. These hours of both machinery and manpower are also quantified in the model, based on previous work by the authors [15]. These management hours for CDWs are included in the budget, separately in its corresponding chapter, as required by Royal Decree 105/2008 [72, 88]. Therefore, for the environmental evaluation of the CDWs, the environmental impact associated with the hours of machinery and manpower will be evaluated in the same way that the evaluation of these resources has been defined in the previous sections.

7.   Direct consumption during the use stage

In general, for the calculation of energy and water consumption, empirical data obtained from simulation software, supply companies, data from other research will be used [89, 90], the Technical Building Code (CTE) [56] and organizations such as the National Statistics Institute [91], the Institute for Energy Diversification and Saving [86] and the World Health Organization (WHO). The parameters used for the estimates of each of the direct consumptions are defined in detail below.

7.1 Water consumption

For the estimation of water consumption, the starting point is the collection of data provided by the National Institute of Statistics on the average water consumption per person per year, from 1996 to 2015. In addition, the optimal consumption value provided by the World Health Organization (WHO) is taken into account, 50 l/inhabitant/day, an amount associated only with human consumption of water in the home (drinking, cooking, personal hygiene and household cleaning). After analysing the evolution of water consumption in recent years, a trend towards savings has been observed and with it, a reduction in water consumption. In the year 2000, average water consumption in Spanish households was above 150 L per person per day, compared to 132 L per person per day in 2018, according to the latest data collected by the National Statistics Institute [91].

Based on this, a polynomial trend curve has been estimated, see Fig. 7, which extends over the years of the BLC. Once the trend reaches the 50 l/inhabitant/day recommended by the WHO, the value remains constant. According to the data obtained, this figure is expected to be reached around 2038. With all of the above, together with the estimated number of residents per dwelling (according to CTE values, 2006), it is possible to obtain a forecast of the water consumption that will occur in the building.

For the evaluation of the environmental impact associated with tap water, it should be remembered that it is not only the quantity of water consumed that should be taken into account, but also the energy expenditure for its transport from the source to the point of consumption, as well as losses due to leaks and breakdowns. In Spain, it is currently estimated that there are 15.9% losses in public urban supply networks, according to a report by the National Institute of Statistics [91]. In addition, the energy associated with the treatment of urban water collection, supply and distribution is estimated at 8.345 kW/m^3 of pumping electricity consumption according to the Institute for Energy Diversification and Saving [86], in its article entitled Water Supply and Treatment. After evaluating all the aspects associated with tap water [92], it is established that the environmental impact in EF for each m^3 is 0.0002 hag/m^3, being the value applied in this research.

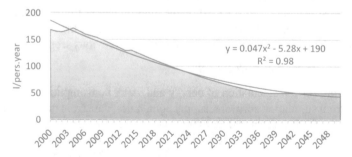

$$y = 0.047x^2 - 5.28x + 190$$
$$R^2 = 0.98$$

Fig. 7 Per capita human water consumption trend (self-made from OMS and [91])

7.2   Electricity consumption

With regard to electrical consumption, the first thing that is needed is an estimate of the consumption that the building will require according to its construction characteristics. To this end, the case studies have been subjected to a simplified consumption analysis through the CE3x_housing energy certification programme (CE3X v2.3). The initial configurations of the buildings have been modelled using the thermal transmittances of their construction solutions according to the construction projects and the subsequent improvement measures defined based on the new transmittances improved in the successive renovation projects, obtaining consumption and emission data.

Something important to take into account given the duration of the BLC are the possible changes in the environmental impact of the energy consumed. The continuous search to minimize the environmental impacts caused by the current lifestyle leads to the progressive improvement of the manufacturing processes of resources, as well as the search for sources of energy generation that reduce GHG emissions as much as possible, that is, energy matrices, tends to increase the presence of renewable sources. For this reason, the model includes the evolution of the energy mix over time. With the combination of data collected from predictions of the energy mix [87] and data obtained from LCA databases [93], a possible future scenario was generated, where the different trends (polynomials) of the energy sources in the Spanish mix were analysed.

According to the data obtained, see Fig. 8, it can be seen how the evolution of energy sources is moving towards the use of renewables, raising the possibility that, around 2070, our energy sources will only be solar, wind and/or hydroelectric. This allows us to estimate the environmental impact produced by each kWh consumed in accordance with the period of use of the building.

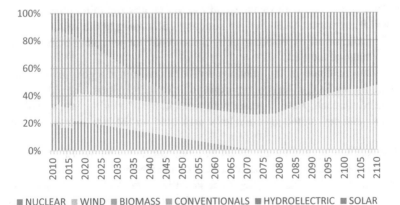

Fig. 8 Proposed future scenario for the evolution of electricity sources in Spain (2010–2110) (self-made from REE, 2018, and [93]

Table 5 Environmental impact of electricity consumption by period (self-made from [87, 93])

Electricity	EF (hag/kWh)
Period 0–20	0.00013
Period 21–40	0.00012
Period 41–70	0.00011
Period 71–100	0.00010

To conclude and facilitate the environmental updates taken into account for the EF indicator, Table 5 shows, by way of summary, the impacts of electricity adapted to the expected future scenario, for the different periods of consumption established in the BLC.

(c)   Creation of the economic and environmental database for BLC

It is essential to create an auxiliary database and systems that allow for the management of the large amount of data from the building project, whether at the level of construction, use and maintenance, or demolition of the building. It is considered necessary to develop a project measurement management system, which organizes and classifies the data of the resources needed during each phase of the BLC. The standard ISO/TR 14177:1994 [94] shall be followed.

In the same way that the resource inventory was used for the application of the EF methodology, in this section, work is being done on the creation of a database called the Resource Quantification Bank (BCRR), which is responsible for breaking down these resources into the categories necessary for calculating environmental impact, materials, machinery, manpower, waste (indirect resources) and water and electricity consumed during the use of the building (direct resources).

The BCRR should be developed independently, but in a coordinated manner for each stage of the BLC, as should be the calculation models for the development, construction, use and maintenance, and renovation or demolition phases. The data obtained from project quantities are structured according to the ACCD systematic classification and are expressed in units of measurement per unit of built area (u/m^2) for each stage of the BLC.

The data obtained in the previous divisions are then merged, giving rise to a database for the BLC of an economic and environmental nature. The unit costs (UC) considered for each phase, together with simulations of building use, will allow the economic and environmental costs to be estimated for the 100-year duration of the BLC.

The ACCD's pyramid-shaped budget base, together with the EF methodology, has enabled the creation of a database that combines economic and environmental information (Fig. 9). The first step to be taken in order to obtain the EF and CDW of each element (or BC) consists of converting the unit of measurement that is the origin of each BC (m^3, m^2, m, t, thousands of units, etc.) a m^3, so that the density established in the support documents used Catalogue of Construction Solutions of the Technical Building Code [95] and the Basic Document on Structural Safety of

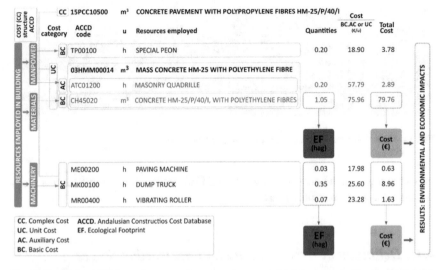

Fig. 9 Example of UC from the economic and environmental database created for the BLC (self-made from [19])

the Technical Building Code. Actions in Building DB-SE AE [56], and thus obtain the weight of each element.

Once the EF has been obtained for each of the materials listed, these are complemented with their "environmental costs" per unit of reference, which will mark the basic environmental impacts. Figure 9 shows an example of a UC, where the Basic EF (BEF) has been incorporated together with its BC. With all these basic impacts created, and following the initial pyramidal structure, the unitary impacts (UEF) are obtained in the same way as the UCs in the ACCD. Finally, if this economic and environmental cost base created is applied to the project measurements, it is possible to calculate, at the same time, the economic costs and the environmental costs (EF) that any building project will involve.

4 Case Study (Model Application)

The case study is a project of a residential complex of houses representative of the most built typologies in Spain between 2006 and 2010 [2]. The urbanization covers 7,123.8 m² plot and corresponds to the work prior to the construction of two blocks of multi-family houses of four floors: one has 57 houses of 88. 81 m² and the other has 50 houses of 89.24 m²—in total 107 homes and 9,524.17 m² floor area. See Fig. 10.

The selected project begins its stages of urbanization and construction in the years 2008–2009, establishing the beginning of the stage of use in 2010. The number of

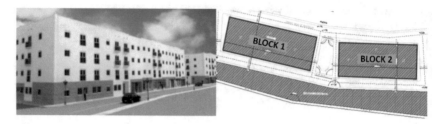

Fig. 10 Case study residential complex (self-made from [14])

people living in each house is set by the CTE [56], setting three people for this study because it is a two-bedroom house.

4.1   Definition of Urbanization and Construction Stages

The following items are included in the urbanization project:

- Cleaning and clearing of the land.
- Roadway works, which included earthworks for the levelling, compacting and paving of roads; construction and taping of floor slabs; and construction of the channels to be built in the subsoil of the tracks or slabs.
- Sewerage works, general and partial collectors, connections to supply networks and drains.
- Works for the installation and operation of public services for water supply, electricity, public lighting, telephony and telecommunications.

For the construction of the houses, a building has been designed with a surface foundation with reinforced slab and reinforced concrete structures, as well as metal formwork systems. For the horizontal structures selected ceramic vault slabs. For the enclosure, the façade is made up of 24 cm brick walls with an air chamber, 3 cm polystyrene insulation and exterior cladding in artificial stone. The roofs are horizontal, passable and ventilated. The interior divisions are executed with 9 cm brick partitions. For the interior finishes, the walls, terrazzo floors and plaster ceilings have been trimmed and plastered with metal fixings. For the holes in the houses, we have chosen lacquered aluminium windows with thermal bridge rupture (TBR), thermo-acoustic glasses 6 + 12 + 6, PVC blinds and without protection grills.

Other finishes to consider are melamine wood doors and steel railings. It has been decided to implement an air conditioning system with a heat pump and console terminal in the facility. For the domestic hot water system, a solar energy system with thermoelectric support is chosen. The water supply pipes are of copper and the drainage pipes are of reinforced PVC. For accessibility, lifts are included. The consumption of resources taken into account for the evaluation of the project is shown in Table 6, by budget chapter and by constructed area.

Table 6 Resource consumption per floor area in case study (self-made)

COD	U	Concept	Quant. (U/m^2)	COD	U	Concept	Quant. (U/m^2)
02E	m^3	Excavations	0.18	08FC	m	ACS conduits	0.16
02R	m^3	Fillings	0.13	08FD	u	Drainage conduits	0.02
02T	m^3	Earth transport	0.22	08FF	m	AFS conduits	0.31
03A	kg	Reinforcement	8.95	08FG	u	Taps	0.02
03P	m	Piles	0.00	08FS	u	Sanitary ware	0.02
03E	m^2	Formwork	0.03	08FT	u	Thermal/heating units	0.01
03HA	m^3	Concrete foundations	0.14	08NA	u	Accumulators	0.01
03HM	m^3	Mass concrete	0.01	08NE	u	Load-bearing structures	0.01
03H	m^3	Strapping concrete	0.00	08NO	u	Solar collectors	0.01
04A	u	Chests	0.01	08NP	m	Primary circuit	0.09
04C	m	Collectors	0.05	09T	m^2	Insulation	0.64
04B	m	Drainpipes and cups	0.11	10AA	m^2	Tiling	0.35
05F	m^2	Forging	0.99	10AC	m^2	Veneer	0.95
05AA	kg	Reinforced	12.67	10CE	m^2	Rendering	1.35
05HE	m^2	Formwork	0.81	10CG	m^2	Garnishes	2.26
05HA	m^3	Reinforced concrete	0.10	10S	m^2	Flooring	0.88
06DC	m^2	Partition walls (chambers)	0.64	10SS	m^2	Concrete floors	0.03
06DT	m^2	Partition walls	0.66	10 T	m^2	Roofing	0.07
06LE	m^2	Brick walls ext	0.95	10R	m	Finishing	0.08
06LI	m^2	Brick walls ext	0.30	11CA	m^2	Steel carpentry	0.11
07H	m^2	Horizontal roof	0.29	11CL	m^2	Light carpentry	0.11
07I	m^2	Inclined roof	0.00	11 M	m^2	Wood carpentry	0.02
08CA	u	Eq. Air conditioning	0.01	11MP	m^2	Doors	0.11
08CC	m	Conduits	0.01	11B	m^2	Handrails	0.06
08CR	m^2	Radiators	0.01	11P	m^2	Blinds	0.06
08EC	m	Circuits	0.53	11R	m^2	Enclosure	0.04
08ED	m	Lines and shunts	0.11	12A	m^2	Glazing	0.10
08EL	u	Light points Power	0.10	13PE	m^2	Exterior paints	1.19
08ET	u	Electrical socket	0.19	13PI	m^2	Interior paints	2.51
08EP	m	Grounding line	0.09	08MA	u	Lifts	0.002

4.2 Definition of Use and Renewals Stages

For the purpose of this research, electricity consumption does not include variations in habits or climate. However, if we take into account the variations caused by the renovations planned in the building, which improve the energy efficiency. To this end, different periods of consumption are established, the beginning and end of which are marked by the different renovation works, the first from the construction of the building to the year 20, the second from the year 21 to 40, the third from 41 to 70 and the last from 71 to 100. In each of these periods, the building's consumption is gradually reduced.

Quantitatively, the improvement parameters applied to renewal actions are shown in Table 7. On the one hand, the thermal transmittance (U) of the enclosure has been improved, both on the roof and on the façade, and in the openings, doors and windows. These improvements will take place in the renovations scheduled at 40 years old. At 70 years old, the same U values are maintained.

On the other hand, in each of the interventions, action will be taken on the DHW production and air conditioning systems. For the characteristics of the air conditioning system, with a heat pump, the values for the nominal performances of the initial heat pumps installed in the building are 220% for heating and 200% for cooling. As the equipment is replaced 20 years later, it is estimated that yields will be improved by industrial advances, using values for the simulations of 350% for heating and 330% for cooling. With regard to the renewal of the DHW installation, the characteristics of the elements that make up the solar panel system for DHW production are maintained. However, the electric thermos supporting the system, covering 30% of the demand, will be improved by installing a thermal storage tank with a capacity of 100 l and thermal insulation. In terms of performance, it is estimated that the DHW installation will remain similar to the initial case.

To obtain the data as close to reality as possible, the energy simulation has been carried out with the official Spanish software CE3x_viviendas (CE3X v2.3) [96] of the case presented, and the various savings that would result from the execution of the aforementioned renovation works have been assessed. These reductions, as well as the annual electricity consumption per m^2, can be seen in a quantitative way in Table 8, where the evolution of consumption in the different periods of use in the BLC is represented.

Table 7 Expected energy efficiency improvements in renovations (self-made)	Improvements	Initial U value (W/m^2 K)	Improved U value (W/m^2 K)
	Facades	0.81	0.57
	Roofing	2.27	0.80
	Frames	4.00	2.20
	Glass	3.30	1.80

Table 8 Consumption in BLC use stage (self-made)

Period (years)	Power consumption (kWh/m²/year)	Energy saving (kWh/m²/year)	Energy rating (consumption)	Emissions (kgCO₂ eq/m²/year)	Emissions savings (kgCO₂ eq/m²/year)	Energy rating (emissions)
0–20	63.18		E	11.84		E
21–40	27.6	−35.58	D	4.89	−6.95	C
41–70	20.95	−6.65	C	3.63	−1.27	C
71–100	8.98	−11.97	B	1.29	−2.34	A

As far as water consumption is concerned, it is estimated that the average consumer has responsible consumption habits. In order to project these consumption habits, the trend towards a reduction in tap water consumption has been taken into account based on the studies set out in the theoretical model, see Fig. 7. The results obtained, reflected in terms of built area, for water consumption in BLC are 68.50 m^3/m^2.

4.3 Demolition Stage Definition

When the 100th year of the building's life is reached and considering that the building does not meet the conditions of habitability, the project for the demolition of the building is carried out. The demolition will be carried out in a massive way with mechanical means. In addition, the management and transport to the treatment plant of all the CDW generated are contemplated.

5 Results and Discuss

The results are presented by differentiating the resources incorporated in the consumption of the works (indirect consumption) from consumption during the use of the building, i.e. water and electricity supplies (direct consumption). All this is evaluated economically and environmentally, and standardized by its corresponding floor area (m^2). Next, in order to facilitate the comparison of the results with other similar studies, the total impacts are presented, in addition to the values affected by the years of duration of the BLC, which, in the case of the present study, is 100. The data applied for the evaluations of one-off activities refer to values for 2018, while data extending over long periods of time, such as supply consumption, are obtained based on the predictions of the future scenarios raised in the model.

5.1 Indirect Consumption Results

Table 9 is presented with a double division of the results. The basic resources for the execution of each stage are broken down, specifically into materials, machinery and manpower, as well as the CDWs generated and the hours of machinery and manpower required to manage them. The resources relating to machinery and manpower are quantified in the total consumed by each of the phases in h/m^2. The total EF data are expressed in global hectares (hag/m^2).

Resources, CDW and environmental impacts by BLC stages (self-made)

		RESOURCES (kg/m²)	CDW (kg/m²)	EF (hag/m²)
Land transformation	Materials (kg)	41.54	310.30	0.03
	Machinery (h)	0.16	0.10	0.01
	Manpower (h)	5.91	0.16	0.002
Building	Materials (kg)	2,159.25	108.32	0.18
	Machinery (h)	0.41	0.06	0.01
	Manpower (h)	10.76	0.06	0.003
Renewal 20	Materials (kg)	7.22	7.26	0.005
	Machinery (h)	0	0.002	0.00001
	Manpower (h)	0.78	0.002	0.0002
Renewal 40	Materials (kg)	659.98	562.63	0.085
	Machinery (h)	0.26	0.18	0.005
	Manpower (h)	6.40	0.19	0.002
Renewal 70	Materials (kg)	1,102.82	1,013.95	0.134
	Machinery (h)	0.41	0.32	0.007
	Manpower (h)	7.85	0.35	0.002
Demolition	Materials (kg)	0	2,159	0
	Machinery (h)	0.04	0.21	0.037
	Manpower (h)	0.001	0.22	0.001

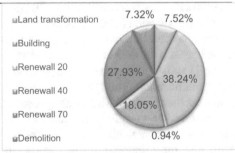

To give a visual idea of the proportion of the impacts, the table is accompanied by a graph. It can be seen that the biggest impacts are those generated by the materials necessary for the execution of the works, and, as is to be expected, the sources of consumption are found in the construction stage.

There is also an important focus on the hours of machinery and manpower required in the renovation programmed in the year 70 of the BLC. This is related to the increase in the CDWs generated by these interventions, since, at this age of the building, the renovation works are due to the repair works of structures. In addition, these results can be related to those presented in the research of Alba et al. [97] in which the economic and environmental profitability of this type of intervention is evaluated in relation to the damage to the building. If these results are analysed in detail, the first thing to be observed is that the consumption of resources in the construction stage is double that is required in the renovation stage 70. However, when we look at the CDWs, we see that they increase almost 10 times more in renovation than those generated in construction. This translates into a notable increase in the impact caused by the management of the CDW generated associated with the machinery.

In the case of demolition, the low impact is more evident, since no material resources are consumed, the only impact associated with this stage of the building being the machinery, both for the execution of the demolition and for the subsequent management of the CDWs generated. It should be remembered that CDWs have no material environmental impact because this was accounted for when the building was incorporated as a new material. When the materials are removed from the building, already considered CDW, only impacts are associated with it because of the machinery necessary for its management and thus, avoiding duplication in the quantification of impacts.

Finally, the stage of renewal at 20 years stands out as the stage of least impact, around 1%. In the case of the renovation after 20 years, these reduced impacts are due to the fact that the actions in the building only intervene in the replacement of the air conditioning and sanitary hot water generation equipment (DHW), works of little entity if compared to the rest of the stages.

The impacts associated with materials and CDWs are discussed in more detail below. To this end, Fig. 11 shows the classification of the main families of materials, the material resources consumed and the CDWs generated throughout the entire BLC. The total environmental impact generated by each type of material has been calculated. The quantification has been carried out by measuring the weight of material consumed per surface area of the house built for each case (kg/m^2).

Figure 11 shows how more CDW is generated than materials consumed in almost all families of materials. This is due to the fact that, in the renovation stages, it must be taken into account that all the materials incorporated will be to replace other damaged ones; therefore, the quantities generated of CDW are quite close to the materials consumed. To these amounts, they must add the losses inherent in the execution of the work and, most importantly, the increase in CDW due to the packaging of the materials, the main reason why the final balance between materials and CDW is greater in quantity than the total CDW in the BLC. Finally, the demolition stage, a phase in which no materials are consumed, only CDW are generated, which are the

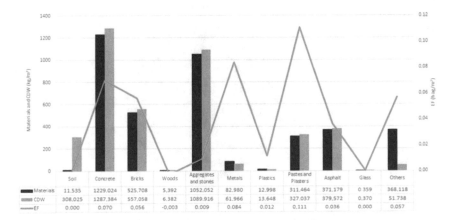

	Soil	Concrete	Bricks	Woods	Aggregates and stones	Metals	Plastics	Pastes and Plasters	Asphalt	Glass	Others
Materials	11.535	1229.024	525.708	5.392	1052.052	82.980	12.998	311.464	371.179	0.359	368.118
CDW	308.025	1287.384	557.058	6.382	1089.916	61.966	13.648	327.037	379.572	0.370	51.738
EF	0.000	0.070	0.056	-0.003	0.009	0.084	0.012	0.111	0.036	0.000	0.057

Fig. 11 Analysis of resources consumed and CDW generated (self-made)

equivalent, in quantity, to the materials quantified in the building's construction phase, discounting the CDW. In other words, all the materials that were once consumed in the construction of the building, once demolished are converted into CDW. The materials that are most represented in the BLC after concrete are aggregates and stones, followed by brick. In contrast, the least representative materials are glass, wood and plastic, in order of lowest to highest consumption in the BLC. The 10 main families of materials identified in the results coincide with those identified by Solís-Guzmán et al. [98] in Spain, in projects in Italy [99], and Chastas [100], Brazil [101] and Chile [102]. By observing the materials and their associated environmental impacts, it can be seen how the greatest impacts are not given by the most abundant materials.

Finally, another option offered by the results presented in Fig. 11 is to analyse the viability of revaluing CDWs to convert them back into construction materials. This would meet the demand for resources, while reducing the deposit of CDWs in landfills, in line with the new legislation on the circular economy.

Figure 12 shows the proportion of the environmental impacts of the indirect consumption of the project studied. The first thing that stands out is that the environmental impact of the materials is much greater than that of machinery and manpower. The impact produced by the food derived from the maintenance of the manpower represents only 1–42% of the total EF; being the rest divided into 20–21% for the machinery; and the remaining impact 78–37% for the materials. We can see how manpower, being such a consumed resource, represents the least of the environmental impacts. This is due to the high impact associated with the fuels or electricity needed to operate the machinery.

The EF indicator is the only one in the footprint family that allows the quantification of the impact associated with human resources. This is thanks to the involvement of the productive land in the methodology of the same. The productive land provides

Fig. 12 EF of material resources, manpower and machinery in the BLC (self-made)

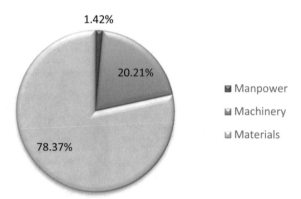

1.42%

20.21%

- Manpower
- Machinery
- Materials

78.37%

the necessary surface area to produce the food that the workers will consume, this being understood as "their fuel" for the work they carry out. This fact defends the use of the EF indicator in the environmental assessment of the FBP, since the industry of the building sector would not be possible without the resource of the manpower force.

5.2 Direct Consumption Results

Focusing on the results shown in Table 10, the first thing to be observed is the impact of indirect consumption on the BLC, which accounts for 60% of the total impact. This places the focus of action on the works, that is, on controlling the building projects. Therefore, if the aim is to reduce the damage that the building sector causes to the environment, this will be achieved in a much more efficient way if the material resources, machinery and manpower of the buildings are controlled instead of only focusing the effort on reducing direct consumption during the stages of use. To all this,

Table 10 Analysis of the direct consumption by stages and total indirect consumption of the BLC (self-made)

		Period (years)	Consumption	EF (hag/m²)
Direct consumption	Water	0–100	68.50 m³/m²	0.01
	Electricity	0–20	1,263.53 kWh/m²	0.16
		20–40	551.99 kWh/m²	0.07
		40–70	628.524 kWh/m²	0.07
		70–100	269.329 kWh/m²	0.03
	Total impact			0.34
Indirect consumption: Materials, machinery and manpower in BLC (Total impact)				0.51

BLC period	Costs (€/m²)	EF (hag/m²)
Land transformation	21.715	0.04
Building	635.29	0.20
Renewal 20	95.93	0.00
Renewal 40	323.12	0.09
Renewal 70	371.31	0.14
Demolition	37.36	0.04
Total works costs	1,484.725	0.51
Electricity consumptions	325.062	0.33
Water consumptions	154.13	0.01
Total consumption of supplies	479.20	0.34
Total in BLC	1,963.93	0.85

Table 11 Direct and indirect consumption cost by BLC stages and projects (self-made)

it should be added that, as the efficiency of new buildings is progressively improved, the weight of the work stage is accentuated, making intervention in building projects more evident in order to minimize the impact on the sector.

5.3 Economic and Environmental Impact Comparative

Below is a summary of the implementation budgets of the different phases of the BLC, as well as their impacts on EFs for the interventions required in each phase of the BLC. As can be seen from the methodology described and the results shown in the comparative table (see Table 11), the budgets are obtained in a way that complements the environmental impacts. This is because these are the resources that are extracted directly from the quantification of the structure of the ACCD, obtaining at the same time the two budgets, the economic and the environmental of the projects evaluated.

The annual consumption starting in 2010 results in a total cost of approximately 1,963.93 €/m². It should be clarified that, although there is the possibility of making economic updates with rates such as those of the CPI in the analyses presented, it has been chosen not to apply any monetary update rate to avoid muddying the analysis of the results.

Figure 13 compares the different budgets in the BLC (economic in figure (a); environmental in figure (b)) for each of the phases, together with the evaluation of consumption in the different periods of use. Going into detail, it can be seen that the economic weight is centred on the costs derived from the work carried out on the building, with the construction stage representing the largest expense, accounting for 32, 35% of the total. Along with construction, renovation at 70 years of age would be the next largest expense of the works, accounting for 18, 91%, whose importance in the budget is justified by the scale of the interventions due to the importance of the damage at that age of the building.

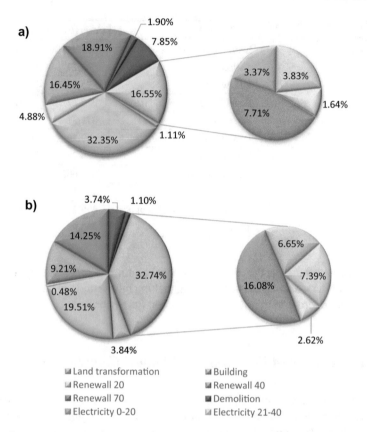

Fig. 13 Comparison of economic impacts (**a**); versus environmental impacts in EF (**b**) (self-made)

Analysing the EF produced by the project in the BLC (see Fig. 13b), indirect consumption is around 60% of the total impact. The rest of the impacts (electricity and water consumption) jointly account for the remaining 40%, with 5% of the total EF, relating to water supply.

In order to have a double perspective of analysis, Fig. 14 shows the evolution of environmental impacts in the BLC, relating the energy impacts represented by the consumption during the programmed works, against those foreseen during the use of the building during its entire useful life.

Another result of interest to analyse is the reduction in expenses associated with consumption. This is because all renovation actions are accompanied by energy improvement of both the building envelope and the installations. In Fig. 14, the relationship between the investment costs of the renovation work planned in the BLC and the reduction in the costs associated with energy consumption can be seen graphically, represented by the yellow line, the annual expenditure on electricity in euros per square metre of built area. The graph also represents the expenses derived

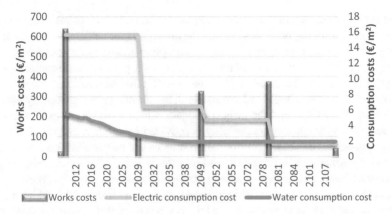

Fig. 14 Evolution of the costs associated with the consumption in the BLC of the projects analysed (self-made)

from water consumption with the blue line (€/m²), where the progressive reduction of these can be seen, in accordance with the scenario proposed for the BLC.

6 Conclusions

After the application of the case study, the analysis obtained has allowed the identification of materials or work units. In addition, it has been possible to obtain the elements that have the greatest impact on the environment, which in the case of materials are metals, followed by concrete, and its derivatives, followed by asphalt and ceramics, the latter being the main elements that control and encourage their reduction and consequently minimize the impact of BLC.

As for the stages of work, those that generate the most pronounced impacts, as expected, are the construction stage and the stage of renovation at the age of 70 of the building. This is due to the high consumption of resources for their execution. With the attention paid to the use phase, the results also show the importance of evaluating the entire BLC since, as can be seen in the results, the impacts are divided 60% in resources during the works and another 40% during use, which implies that not only should our efforts be focused on energy efficiency and savings in consumption in the use phase, but also that the manufacture and use of the resources required in the works should be optimized so that our objectives of reducing the impacts on the construction sector are met.

This research has revealed that, as time goes by, the changes caused by the increase of renewable sources in the energy matrix can minimize the impact of GHG emissions to the atmosphere from the resources consumed. At present, many researches are focused on controlling the CO_2 that is emitted, but the trend of the energy mix reveals that the importance of energy must be valued, since it will always be necessary. This need for energy consumption in the manufacture of materials makes it appropriate to work on improving processes, so that they become more efficient not only at the energy level, but also in terms of water consumption and/or waste generation.

After observing the results obtained, one is in a position to formulate model proposals to include the assessment of the sustainability of a building project. This proposal is based on simple and accessible data. Simplicity is provided by the cost base used, the ACCD and the project budgets, which have served as a vector for introducing sustainability; accessibility is provided by the fact that the data come from open-access databases or information sources and can be consulted by anyone, such as the generic LCA databases; and robustness is provided by the reliability of the sources, which have been intensively tested in numerous investigations.

This has generated a robust and replicable model that offers results, with a double economic and environmental aspect. In this way, it is possible to make decisions from the project phase that will result in the improvement of the "environmental qualification" of the building throughout its useful life, without forgetting the economic aspect.

Acknowledgements We would like to thank the University of Seville for financing the research work presented, through a pre-doctoral contract, for the development of own I+D+i.

References

1. Fundación General De La Universidad Complutense De Madrid. Cambio Global España 2020/50 (2010) Sector edificación. Programa Cambio Global España 2020/50 del Centro Complutense de Estudios e Información Medioambiental. Madrid. España
2. González-Vallejo P, Marrero M, Solís-Guzmán J (2015) The ecological footprint of dwelling construction in Spain. Ecol Ind 52:75–84. https://doi.org/10.1016/j.ecolind.2014.11.016
3. Weidema BP, Thrane M, Christensen P et al (2008) Carbon footprint: a catalyst for life cycle assessment? J Ind Ecol 12:3–6. https://doi.org/10.1111/j.1530-9290.2008.00005.x
4. Bare JC, Hofstetter P, Pennington DW, Haes HAU (2000) Midpoints versus endpoints: the sacrifices and benefits. Int J Life Cycle Assess 5:319–326. https://doi.org/10.1007/BF0297 8665
5. European Parliament—Council of the European Union (2003) Integrated product policy. Development of the environmental life cycle concept, COM, Brussels, Belgium
6. Galli A, Wiedmann T, Ercin E, Knoblauch D, Ewing B, Giljum S (2012) La integración de la familia de indicadores de huella ecológica, carbono e hídrica: Definición y papel en el seguimiento de la presión humana sobre el planeta. Ecol Ind 16:100–112
7. Wackernagel M, Rees W (1996) Our ecological footprint: reducing human impact on the Earth. New Society Publishers, Gabriola Island, BC
8. Finnveden G, Hauschild MZ, Ekvall T, Guinée J, Heijungs R, Hellweg S, Koehler A, Pennington D, Suh S (2009) Recent developments in life cycle assessment. J Environ Manage 91:1–21
9. Herva M, Franco A, Ferreiro S, Álvarez A, Roca E (2008) Un enfoque para la aplicación de la huella ecológica como indicador ambiental en el sector textil. J Hazard Mater 156:478–487
10. Borucke M, Moore D, Cranston G, Gracey K, Katsunori I, Larson J, Lazarus E, Morales JC, Wackernagel M, Galli A (2013) Accounting for demand and supply of the biosphere's regenerative capacity: the National Footprint Accounts' underlying methodology and framework. Ecol Ind 24:518–533

11. GFN (Global Footprint Network) (2014) Paquete de aprendizaje de la Huella Nacional.Disponible: https://www.footprintnetwork.org /en/index.php/GFN/page/licenses1/ (visitada 01/01/16)
12. Giljum S, Burger E, Hinterberger F, Lutter S, Bruckner M (2011) Un amplio conjunto de indicadores de uso de recursos micro al nivel macro. Resour Conserv Recycl 55:300–308
13. Solís-Guzmán J, Marrero M (2014) Ecological footprint assessment of building construction. Bentham Sci Pub 2015, 162 (2014)
14. Solís-Guzmán J, Marrero M, Ramírez-De-Arellano A (2013) Methodology for determining the ecological footprint of the construction of residential buildings in Andalusia (Spain). Ecol Indic 25. doi:https://doi.org/10.1016/j.ecolind.2012.10.008
15. Marrero M, Puerto M, Rivero-Camacho C et al (2017) Assessing the economic impact and ecological footprint of construction and demolition waste during the urbanization of rural land. Resour Conserv Recycl 117. doi:https://doi.org/10.1016/j.resconrec.2016.10.020
16. González-Vallejo P, Solís-Guzmán J, Llácer R, Marrero M (2015) La construcción de edificios residenciales en España en el período 2007–2010 y su impacto según el indicador Huella Ecológica. Informes De La Construcción 67(539):e111. https://doi.org/10.3989/ic.14.017
17. Martínez-Rocamora A, Solís-Guzmán J, Marrero M (2016) Toward the ecological footprint of the use and maintenance phase of buildings: utility consumption and cleaning tasks. Ecol Indic 69:66–77. https://doi.org/10.1016/j.ecolind.2016.04.007
18. Alba-Rodríguez MD, Martínez-Rocamora A, González-Vallejo P et al (2017) Building rehabilitation versus demolition and new construction: economic and environmental assessment. Environ Impact Assess Rev 66:115–126. https://doi.org/10.1016/J.EIAR.2017.06.002
19. Freire Guerrero A, Alba Rodríguez MD, Marrero M (2019) Madelyn: a budget for the ecological footprint of buildings is possible: a case study using the dwelling construction cost database of Andalusia. En: Sustainable Cities and Society, vol 51. https://doi.org/https://doi.org/10.1016/j.scs.2019.101737
20. UNE-EN ISO 14040 (2006) Environmental management—life cycle assessment—principles and framework
21. UNE-EN ISO 14044 (2006) Environmental management—life cycle assessment—requirements and guidelines
22. Knoeri C, Sanyé-Mengual E, Althaus H-J (2013) LCA comparativo de hormigón reciclado y convencional para aplicaciones estructurales En t. J. Evaluación del ciclo de vida 909–918. https://doi.org/10.1007/s11367-012-0544-2
23. Zabalza BI, Valero CA, Aranda UA (2011) Life cycle assessment of building materials: comparative analysis of energy and environmental impacts and evaluation of the eco-efficiency improvement potential. Build Environ 46:1133–1140
24. Guggemos AA, Horvath A (2005) Comparación de los efectos ambientales de los edificios con estructura de acero y hormigón. J Infraestructura Syst 11:93–101. https://doi.org/10.1061/(ASCE) desde 1.076 hasta 0.342 (2005) 11:2(93)
25. Kua HW, Maghimai M (2017) El debate sobre el acero contra el hormigón revisado: potencial de calentamiento global y análisis de energía incorporados basados en perspectivas atribucionales y consecuentes del ciclo de vida J Ind Ecol 21:82–100. https://doi.org/10.1111/jiec.12409
26. Blengini GA, Di Carlo T (2010) The changing role of life cycle phases, subsystems and materials in the LCA of low energy buildings. Energy Build 42:869–880
27. Verbeeck G, Gallinas H (2010) Inventario del ciclo de vida de los edificios: un método de cálculo. Construir. Reinar. Tiempos 45(2010):1037–1041. https://doi.org/10.1016/j.buildenv.2009.10.012
28. Skaar C, Labonnote N, Gradeci K (2018) De edificios de cero emisiones (ZEB) a vecindarios de cero emisiones (ZEN): una revisión de mapeo de LCA basado en algoritmos Sostener. Times 10. https://doi.org/10.3390/su10072405
29. Trigaux D, Oosterbosch B, De Troyer F, Allacker K (2017) Una herramienta de diseño para evaluar la demanda de energía para calefacción y el impacto financiero y ambiental asociado en los vecindarios Construcción de energía. 152:516–523. https://doi.org/10.1016/j.enbuild.2017.07.057

30. Buyle M, Braet J, Audenaert A (2013) Life cycle assessment in the construction sector: a review. Renew Sustain Energy Rev 26:379–388. https://doi.org/10.1016/j.rser.2013.05.001
31. Ramesh T, Prakash R, Shukla KK (2010) Life cycle energy analysis of buildings: an overview. Energy Build 42:1592–1600. https://doi.org/10.1016/j.enbuild.2010.05.007
32. Schwartz Y, Raslan R, Mumovic D (2018) The life cycle carbon footprint of refurbished and new buildings—a systematic review of case studies. Renew Sustain Energy Rev 81:231–241. https://doi.org/10.1016/j.rser.2017.07.061
33. Chau CK, Leung TM, Ng WY (2015) A review on life cycle assessment, life cycle energy assessment and life cycle carbon emissions assessment on buildings. Appl Energy 143:395–413. https://doi.org/10.1016/j.apenergy.2015.01.023
34. Cabeza LF, Rincón L, Vilariño V et al (2014) Life cycle assessment (LCA) and life cycle energy analysis (LCEA) of buildings and the building sector: a review. Renew Sustain Energy Rev 29:394–416. https://doi.org/10.1016/j.rser.2013.08.037
35. Asif M, Muneer T, Kelley R (2007) Life cycle assessment: a case study of a dwelling home in Scotland. Build Environ 42:1391–1394. https://doi.org/10.1016/j.buildenv.2005.11.023
36. Dimoudi A, Tompa C (2008) Energy and environmental indicators related to construction of office buildings. Resour Conserv Recycl 53:86–95. https://doi.org/10.1016/j.resconrec.2008.09.008
37. Sartori I, Hestnes AG (2007) Energy use in the life cycle of conventional and low-energy buildings: a review article. Energy Build 39:249–257
38. Sharma A, Saxena A, Sethi M, Shree V (2011) Life cycle assessment of buildings: a review. Renew Sustain Energy Rev 15:871–875
39. Samadpour P, Faryadi S (2008) Determination of ecological footprints of dense and high-rise districts, case study of Elahie neighbourhood, Tehran. J Environ Stud 34(45):63–72
40. Zhao XY, Mao XW (2013) Comparison environmental impact of the peasant household in han, zang and hui nationality region: Case of zhangye, Gannan and Linxia in Gansu Province. Shengtai Xuebao/Acta Ecologica Sinica 33(17):5397–5406
41. Li B, Cheng DJ (2010) Hotel ecological footprint model: Its construction and application. Chin J Ecol 29(7):1463–1468
42. Bin G, Parker P (2012) Measuring buildings for sustainability: comparing the initial and retrofit ecological footprint of a century home—THE REEP House. Appl Energy 93:24–32
43. Olgyay V (2008) Greenfoot: a tool for estimating the carbon and ecological footprint of buildings. American Solar Energy Society—SOLAR 2008. Including Proc. of 37th ASES Annual Conf., 33rd National Passive Solar Conf., 3rd Renewable Energy Policy and Marketing Conf.: Catch the Clean Energy Wave, vol 8, pp 5058–5062
44. Teng J, Wu X (2014) Eco-footprint-based life-cycle eco-efficiency assessment of building projects. Ecol Ind 39:160–168
45. Bastianoni S, Galli A, Pulselli RM, Niccolucci V (2007) Environmental and economic evaluation of natural capital appropriation through building construction: practical case study in the Italian context. Ambio 36(7):559–565
46. Radhi H, Sharples S, Fikiry F (2013) Will multi-facade systems reduce cooling energy in fully glazed buildings? A scoping study of UAE buildings. Energy Build 56:179–188. https://doi.org/10.1016/j.enbuild.2012.08.030
47. You T, Stansfield I, Romano MC, Brown AJ, Coghill GM (2011) Analizando el control de traslación GCN4 en levaduras mediante modelación y simulación cinética química estocástica. BMC Syst Biol 5:131
48. Cellura M, Guarino F, Longo S, Mistretta M (2014) Energy life-cycle approach in Net zero energy buildings balance: operation and embodied energy of an Italian case study, Energy and Buildings, Volume 72, 2014. ISSN 371–381:0378–7788. https://doi.org/10.1016/j.enbuild.2013.12.046
49. Scheuer C, Keoleian GA, Reppe P (2003) Life cycle energy and environmental performance of a new university building: modeling challenges and design implications. Energy Build 35:1049–1064

50. SpainGBC (2017) LEED Certificate. https://www.spaingbc.org/web/leed-4.php. Accessed 30 Nov 2018
51. BREEAM (2017) BREEAM ES website. https://www.breeam.es/. Accessed 30 Dec 2018
52. Adalberth K (1997) Energy use during the life cycle of buildings: a method. Build Environ 32(4):317–320
53. UNE-EN 15804 (2012) Sustainability of construction works—environmental product declarations—core rules for the product category of construction products
54. UNE-EN 15459-1:2018. AENOR, 2018. Eficiencia energética de los edificios. Procedimiento de evaluación económica de los sistemas energéticos de los edificios. Parte 1: Método de cálculo, Módulo M1-14. Asociación Española de Normalización y Certificación. Madrid, España (2018)
55. ISO (2011) ISO-15686–1:2011. International Organization for Standardization, ISO. Buildings and constructed assets—service life planning—part 1: general principles and framework. Switzerland: ISO
56. Royal Degree 314/2006, de 17 de marzo, por el que se por el que se aprueba el Código Técnico de la Edificación. Texto refundido con modificaciones del RD 1371/2007, de 19 de octubre, y corrección de errores del BOE de 25 de enero de 2008. España. Revisión vigente desde 13 de Septiembre de 2013. Boletín Oficial del Estado, 28 de marzo de 2006, núm. 74, pp.11816–11831 [Consultado 16 abril 2015]. Disponible en: https://www.boe.es/boe/dias/2006/03/28/pdfs/A11816-11831.pdf
57. Larralde L (2014) Evaluación de la Huella Ecológica de la edificación en el sector residencial de Méjico. Universidad de Sevilla, España, Trabajo Fin de Máster
58. UNE-EN 15978 (2012) Sustainability of construction works. Assessment of environmental performance of buildings. Calculation Method
59. Martínez Rocamora A, Solís-Guzmán J, Marrero M (2017) Ecological footprint of the use and maintenance phase of buildings: maintenance tasks and final results. Energy Build 155. doi:https://doi.org/10.1016/j.enbuild.2017.09.038
60. Freire-Guerrero A, Marrero-Meléndez M (2015) Ecological footprint in indirect costs of construction, pp 969–980. In: Proceedings of the II International congress on sustainable construction and eco-efficient solutions, Seville, 25–27 May 2015
61. CSI/CSC (2016) Construction Specifications Institute/Construction Specifications Canada. Masterformat Manual of Practice (MP2–1). Alexandria, Va
62. UniFormatTM. The Construction Specifications Institute (1998) A uniform classification of construction systems and assemblies. Alexandria, VA
63. Telford T (1991) Civil engineering standard method of measurement. 3rd edn. LTD., UK, pp 4–39
64. Jones AR (1987) CI/SfB construction indexing manual. Royal Institute of British Architects RIBA, London, UK
65. Omniclass (2012) A strategy for classifying the built environment—table 13: spaces by function
66. ITeC (2012) Institute of Construction Technology of Catalonia. Barcelona
67. Official College of Quantity Surveyors, Technical Architects and Building Engineers of Guadalajara (2012) 'PRECIOCENTRO of Guadalajara'
68. Ministry of the Environment and Planning of the Territory, Community of Madrid (2007). 'BPCM Madrid'
69. Department of Housing, Public Works and Transport of the Basque Government (2012) 'BDEU in the Basque Country'
70. Ministry of Infrastructure, Territory y Environment (2014) BDC-IVE Valencia. https://www.five.es/basedatos/Visualizador/Base14/index.htm
71. Andalusia Government (2018) Andalusia construction cost database (ACCD). https://www.juntadeandalucia.es/organismos/fomentoyvivienda/areas/vivienda-rehabilitacion/planes-instrumentos/paginas/ACCD-sept-2017.html. Accessed 11 Jan 2019
72. Marrero M, Ramirez-De-Arellano A (2010) The building cost system in Andalusia: application to construction and demolition waste management. Constr Manage Econ 28(5):495–507

73. Ramírez-de-Arellano-Agudo A (2010) 'Presupuestación de Obras'. Editado Por El Secretariado de La Universidad de Sevilla (1998). Sevilla
74. Freire Guerrero A (2017) Presupuesto Ambiental. Evalución de la Huella Ecológica del Proyecto a Través de la Clasificación de la Base de Costes de la Construcción de Andalucía. Tesis Doctoral
75. Martínez-Rocamora A, Solís-Guzmán J, Marrero M (2016) LCA databases focused on construction materials: a review. Renew Sustain Energy Rev 58:565–573. https://doi.org/10.1016/j.rser.2015.12.243
76. ECO-Platform (www.eco-platform.org/)
77. Ecoinvent Centre (2013) Ecoinvent database v3 ecoinvent report. www.ecoinvent.org. Accessed 30 Dec 2018
78. Frischknecht R, Jungbluth N, Althaus H-J, Doka G, Dones R, Heck T, Hellweg S, Hischier R, Nemecek T, Rebitzer G (2005) The ecoinvent database: overview and methodological framework (7 pp). Int J Life Cycle Assess 10(1):3–9
79. PRé Sustainability (2016) SimaPro 8. https://simapro.com/. Accessed 28 Mar 2018
80. Solís-Guzmán J (2011) Evaluación de la huella ecológica del sector edificación (uso residencial) en la comunidad andaluza. (Assessing the ecological footprint of the building sector (residential use) in Andalusia). Universidad de Sevilla, Sevilla
81. Almasi AM, Milios L (2013) Municipal waste management in Spain
82. EMASESA (2005) La sostenibilidad y la gestión. Cómo estábamos, cómo somos. 1975-2005 Disponible: http://www.emasesa.com/wp-content/uploads/2014/03/Informe-de-Sostenibilidad-2005.pdf
83. World Wildlife Fund (WWF) (2014) Página web española de la World Wildlife Fund. http://www.wwf.es/ (acceso 26.05.2014).
84. González Vallejo P, Muñoz Sanguinetti C, Marrero Meléndez, M (2019) Environmental and economic assessment of dwelling construction in Spain and Chile. A comparative analysis of two representative case studies. J Cleaner Prod 208:621–635. https://doi.org/https://doi.org/10.1016/j.jclepro.2018.10.063
85. SEOPAN (2008) Machinery costs manual (in Spanish: Manual de costes de maquinaria). Available:https://www.concretonline.com/pdf/07construcciones/art_tec/SeopanManualCostes.pdf (accessed 01.07.16). [WWW Document]
86. IDAE (2011) Instituto para la Diversificación y Ahorro de Energía: Factores de Emisión de CO_2/CO_2 Emission Factors
87. REE (2014–2018) El Sistema Eléctrico Español/The Spanish Electric System
88. Ministry of Presilence (2008) Real Decreto 105/2008, de 1 de febrero, por el que se regula la producción y gestión de los residuos de construcción y demolición. Diario oficial boletín oficial del estado, n. 38. España, 2008. Disponible en: https://www.boe.es/buscar/pdf/2008/BOE-A-2008-2486-consolidado.pdf
89. Cubillo F, Moreno T, Ortega S (2008) Microcomponentes y factores explicativos del consumo doméstico de agua en la Comunidad de Madrid. N°4. Cuadernos de I+D+i. Canal de Isabel II. Madrid
90. Naredo Pérez JM (coord) (2009) El agua virtual y la huella hidrológica en la Comunidad de Madrid. Cuadernos de I+D+I, Cnal de Isabel II, Madrid
91. INE (2019) Instituto Nacional de Estadística. [Fichero de datos]. Recuperado de https://www.ine.es
92. Botto S (2009) Tap water vs. bottled water in a footprint integrated approach. Nat Prec. https://doi.org/https://doi.org/10.1038/npre.2009.3407.1
93. Rodríguez de Lucio A, Perea B, Larrea P, Sevilla J, Falkner A (2010) El Modelo Eléctrico Español en 2030. Escenarios y Alternativas
94. ISO (1994) ISO/TR 14177. International Organization for Standardization, ISO. Classification of information in the construction industry. ISO/TR 14177:1994. Switzerland: ISO
95. IETcc (2010) Catálogo de Elementos Constructivos Del CTE. Instituto Eduardo Torroja de Ciencias de La Construcción (IETcc). Retrieved https://www.codigotecnico.org/web/recursos/aplicaciones/contenido/texto_0012.html

96. CE3X (2012) CE3X v2.3. - Programa de certificación energética de edificios existentes. Guía IDAE: Manual de usuario de calificación energética de edificios existentes CE3 X

97. Alba-Rodríguez MD, Marrero M, Solís-Guzmán J (2013) Economic and environmental viability of building recovery in seville (Spain). Phase 1: Database in Argis. Srodowisko Mieszkaniowe 11:297–302

98. Solís-Guzmán J, Rivero-Camacho C, Alba-Rodríguez D, Martínez-Rocamora A (2018) Carbon footprint estimation tool for residential buildings for non-specialized users: OERCO2 project. Sustain 10. doi:https://doi.org/10.3390/su10051359

99. Blengini GA (2009) Life cycle of buildings, demolition and recycling potential: a case study in Turin, Italy. Build Environ 44:319–330

100. Chastas P, Theodosiou T, Kontoleon KJ, Bikas D (2018) Normalising and assessing carbon emissions in the building sector: a review on the embodied CO_2 emissions of residential buildings. Build Environ 130:212–226

101. Maciel T, Stumpf M, Kern A (2016) Management system proposal for planning and controlling construction waste. Propuesta de un sistema de planificación y control de residuos en la construcción. Rev. Ing. construcción 31:105–116 https://doi.org/10.4067/S0718-507320160 0200004

102. Rivero Camacho C, Muñoz Sanguinetti C, Marrero Meléndez M (2018) Cálculo de la Huella Ecológica en el ciclo de vida para la fase de urbanización de un conjunto habitacional en Chile, bajo el modelo ARDITEC. Calculation of the Ecological Footprint in the life cycle for the urbanization phase of a housing complex in Chile. In: 2018, I. 2017 ediciones IA (ed) Congreso Interdisciplinario de Investigación En Arquitectura, Diseño, Ciudad y Territorio, Chile, pp 82–99

Ecological Footprint Assessment and Its Reduction for Packaging Industry

Vishal Nawandar, Dilawar Husain, and Ravi Prakash

Abstract Packaging of products is unavoidable as it is necessary to keep the products safe and to carry them easily during their transport and distribution. The paper packaging products are an important part of the overall packaging industry in India. The growth in the Indian paper packaging industry has been largely driven by the enhanced demand for transportation and distribution of several products. Generally, the paper used in the corrugation industry is the recycled kraft paper, while the virgin paper is used in food packaging where food comes in direct contact with the carton or paper. As some kind of packaging is unavoidable, it is expected to be of low environmental impact as well as low cost for both environmental and economic benefits. This study is dedicated to assess the environmental impact of the manufacture of different types of paper packaging corrugated sheets and boxes (two ply, three ply, five ply and seven ply) in terms of Ecological Footprint (EF) as well as to suggest economically viable measures for its reduction through energy efficiency improvements. This study was carried out for an existing packaging industry M/S Vishal Packaging located at Khamgaon, Maharashtra, India. A detailed survey of the industry was carried out for relevant data collection in order to evaluate the associated EF of the corrugated sheets and boxes being produced. This study also suggests some sustainable measures that have the potential to reduce EF as well as production costs for the industry. The EF of two ply, three ply, five ply, and seven ply sheet boxes have been estimated as 0.515, 0.537, 0.527 and 0.524 gha/ton, respectively, for the existing mode. In this study, three different modes of sustainable measures for EF reduction were examined: (1) grid-connected solar PV system for all energy needs,

V. Nawandar · R. Prakash
Department of Mechanical Engineering, Motilal Nehru National Institute of Technology,
Allahabad, UP, India
e-mail: vishalnawandar@gmail.com

R. Prakash
e-mail: rprakash234@gmail.com

D. Husain (✉)
Department of Mechanical Engineering, Maulana Mukhtar Ahmad Nadvi Technical Campus,
Malegaon, India
e-mail: dilawar4friend@gmail.com

© The Author(s), under exclusive license to Springer Nature Singapore Pte Ltd. 2021
S. S. Muthu (ed.), *Assessment of Ecological Footprints*,
Environmental Footprints and Eco-design of Products and Processes,
https://doi.org/10.1007/978-981-16-0096-8_2

(2) LPG fuelled corrugation heater with conventional grid electricity for other energy needs and (3) LPG fuelled corrugation heater with grid-connected solar PV system for other energy needs. By adopting such measures, the potential reduction in EF for Mode 1, Mode 2, and Mode 3 was estimated at approximately 12.42%, 6.25%, and 11.13% of the total EF, respectively. The EF of raw materials for existing mode and the three energy modes are 88.2% (existing mode), 96.99% (Mode 1), 92.59% (Mode 2) and 96.04% (Mode 3) of the total EF. The investment required for the above three modes was also examined and economic payback period for Mode 1, Mode 2, and Mode 3 was evaluated as 5.3 years, 1.2 years, and 4.3 years, respectively.

Keywords Paper packaging · Ecological footprint · Environmental assessment · Economic assessment · Solar PV system

1 Introduction

The natural capital is prerequisite for the manufacture of human-made products, while the opposite is not [22]. This simply explains if natural resource exploitation rate exceeds the regeneration rate of nature, the effects like global warming, increased draughts, rising sea levels, etc. are visible. The ecological footprint can measure the impact of utilization of resources, and it also gives a check on the load created by human settlements on the planet.

To get a fish served in our dish, it is not just the fish stock but also the fishing boats, canning process, cooking energy, etc. are required which altogether leave a more significant impact as compared to a fish which we consume [22]. Thus, our habits of processing food and goods, before use, have led to massive impact on nature by processing industries and transportation. To stay within the limit, the concept of sustainability arose, which in simple words means to remain indefinitely. Sustainable living ensures that we consume the natural resources at the rate less than or equal to the rate of regeneration of those resources. Now to measure sustainable living is not that easy as the extent of utilization of various resources in various combinations and forms is complex and keeping a track on these resources becomes difficult.

With the economic and technological resources of the present day, it is possible to achieve sustainability by strictly adhering to the practices dictated by sustainable development. Sustainable development is defined as the development that meets the needs of the present without compromising the ability of future generations to meet their own needs [22]. Sustainable development is based on three pillars: (1) social, (2) economic and (3) ecological factors.

Various indicators have been developed to measure sustainability. Some indicators are meant to measure a particular resource like Water Footprint (WF), some are used to measure the environmental factors like Carbon Footprint (CF) and Ecological Footprint (EF). Some other indices are used to measure the social factors like Human Development Index (HDI), Living Planet Index (LPI), while some economic indices might include Gross Domestic Product (GDP), Gross National Product (GNP), etc.

Many other indices are used to compare various processes or activities or products based on the emissions, energy required, etc. But to really measure whether the human activities are really sustainable for humans, as well as other living beings on the planet, is a challenging task.

Sustainability is a very broad concept which includes all the human activities and their impacts. Industrial sustainability is the approach which aims to improve the productivity of industries by considering the energy and resources (human and material) used and their impacts on the environment. Industrial sustainability firstly includes financial stability, which is the primary requirement for any industry to sustain. Later it also involves the social and environmental factors which are equally important in the world, which is presently developing very fast and at the same time facing the severe environmental problems like global warming. Industrial sustainability is generally attained by improving technology, improving energy efficiency and switching to renewable energy resources, etc.

Initially, sustainability was thought to forecast the need for industrial products or processes demanded by the market and the raw material and other resources' availability from an economic perspective. Various studies have suggested various tools to measure industrial sustainability in recent times. Ford developed the Product Sustainability Index (PSI), which was based on eight factors of environmental, social and economic importance [19]. Another methodology suggests product life cycle analysis which involves four phases of product: pre-manufacturing, manufacturing, use and post-use; and then measure the suggested Product Sustainability Index (ProdSI) to compare [20].

These product-based indicators are meant to compare the products of similar kinds and justify which product is more sustainable in a given condition. To measure and compare the sustainability of an industry, the Industrial Sustainability Index (ISI) was developed based on the multi-attribute decision theory methodology, which monitored various attributes under three general criteria: financial (industry must exist for long to be sustainable), in-walls (within the industry) and out-walls (area outside the factory but influenced by industry) [3]. This indicator has the flexibility to give importance to general attributes by using weighting fraction and might give varying results based on the varying weighing factor, which is generally based on individual experiences.

Another Industrial Sustainability Index (ISI) was suggested to compare various industries (similar or dissimilar) based on three major factors of sustainability, i.e. economic, social and environmental [14]. The simplicity of this indicator enabled comparing the process/manufacturing industries rather than product-based analyses as suggested by initial indicators. This ISI is a simplified tool, which represents the socio-economic benefit of any type of industry per unit of its carbon emissions. Carbon emissions could be direct and indirect. The direct carbon emissions may be due to the inherent nature of the production process, while the indirect emissions are caused due to energy consumption in the production process. The ISI as proposed assesses social, economic and environmental goals of any type or types of industries (i.e. small, medium or large scale). The concept of ISI is illustrated in Fig. 1. The expression of the ISI is as follows:

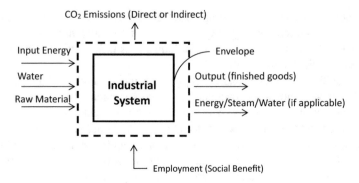

Fig. 1 Sustainability analysis for an industrial system

$$ISI = \frac{(RVA) \times (EMP)}{CO_2 \, \text{emissions}} \qquad (1)$$

Here, the term "RVA" represents the resource value addition (i.e. the difference of the total annual economic values of material, water and energy outputs [products] and that of inputs); it is represented here as million Rs per year. The limitation of RVA using Indian currency (Rs) can be overcome if the RVA is expressed in US dollars with purchasing power parity (i.e. PPP $). The use of purchasing power parity can make the RVA units universal in nature rather than being country-specific. The term "EMP" represents the total number of persons employed by the industry in a year. This refers to manpower with full-time employment (FTE, i.e. 8 h per day). The manpower employed for less than 8 h per day can be proportionately converted to the equivalent FTE. For example, employment for 4 h per day would be equivalent to 0.5 FTE. The term "CO_2 emissions" represents the total annual carbon dioxide emissions (direct and indirect) by the industry during its production process (in tCO_2 per year). Thus, the ISI value is expressed in terms of million Rs* persons per tCO_2 emissions, henceforth expressed as "units" [15].

All these indicators are used to compare the sustainability level of two industries, similar or dissimilar, and also some might be useful to compare the impact of improved industrial practices by the previous ones. But, none of these indicators gives the actual impact of any industry on the environment. The measurement of natural resources exploitation rate is expressed in terms of ecological footprint. This chapter will assess the industry by Ecological Footprint (EF) to estimate the actual environmental impact caused by an industry as well as the Industrial Sustainability Index (ISI as defined in Eq. 1) to check whether the suggested modifications actually improve the sustainability of the industry overall.

After the Industrial Revolution in the nineteenth century, economic and industrial developments have occurred drastically. With that has increased the use of energy utilized for industrial and domestic purposes. The energy is mainly obtained by fossil fuels like petroleum products, natural gas, nuclear resources and biomass. The primary energy and electricity requirement of the world has increased from 71,013 to

Fig. 2 Sector-wise primary
energy utilization of world
[1]

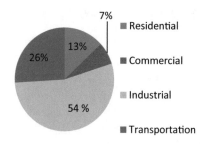

157,482 TWh and 6129 to 23,322 TWh from 1973 to 2014. India's primary energy consumption and electricity consumption for the year 2013 were 9018 TWh (i.e. 12.7% of the world) and 979 TWh (i.e. 4.12% of the world), respectively [1].

Industrial energy consumption was about 54% of the total energy consumption in 2012, as shown in Fig. 2. The major sources of energy for industrial purposes are fossil fuels, mainly coal, natural gas and petroleum products. In recent years, developed countries have stopped manufacturing of goods due to increased daily wages. This has led to industrial developments in developing nations. China is considered as the manufacturing hub of the world, and so it holds maximum industrial energy consumption share. Since the last four decades, China has fostered industrial growth, and this has led to make it the fastest growing economy at present. But with recent tensions and developments, many international companies are trying to shift their manufacturing facilities to developing countries like India, Indonesia, Vietnam, etc. The industrial demands of China may decline by 2035 as predicted by the report [23].

India's energy consumption has been majorly dependent on fossil fuels. In 2017, the primary energy consumption share of coal and petroleum products accounted for 85% of total energy consumption, as depicted in Fig. 2. A few studies suggest that by 2040, coal will still be the major driving source of India, accompanied by natural gas and lowered petroleum dependence [13].

In the coming years, India will be the one of the fastest developing economies, which would become the industrial hub due to cheap labour and abundance of available natural resources. With this, the energy demand in India will rise rapidly, and so the provision for making the energy available for these industrial practices needs to be figured out. In 2004, coal and oil constituted for 60% of industrial energy while electricity use was limited to 14% of total industrial energy consumption.

Industries in India are widely spread according to variable energy-efficient practices. Unlike energy-efficient cement industries, paper and pulp industries are inefficiently handled, utilizing about 70% more thermal and 7% more electricity of developed counterparts [7]. Thus, the paper-based industries require a lot of attention over energy-efficient handling. Five per cent (5%) of the total industrial production accounts for paper and pulp production. Indian paper industry accounts for 1.7% of total paper manufactured in the world [14]. In India, the paper is not manufactured

Fig. 3 Sources of primary
energy in India [1]

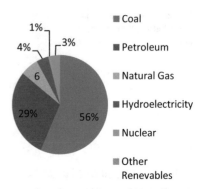

from pulp on a large scale because of heavy dependence on wood for all the household heating purposes. About 25% of paper is manufactured from bamboo, while
58% is recycled (Fig. 3).

Since the paper industry in India is inefficiently handled, more advancement is
essential to arrest the excess energy utilization. Majorly paper in India is recycled
for uses like newspapers, carry bags and cartons. Virgin paper is mandatory for
packaging of all the edible products while the packaging of other products and other
uses is done by recycled paper.

Packaging plays an important role in the transportation of goods. It is an unwanted
affair as it is of no use to the consumer, but unavoidable for safe transportation as well.
However, perishables need leak-proof packing, and expensive goods require hard
packing for safe handling. This chapter focuses on the impact of a paper corrugation-
based packaging industry.

Kraft paper and paper board are the main raw material of the corrugated box.
The packaging paper and paper board consumption in India during the financial year
2016 was 7.9 million tonnes which were about 49% of the total paper requirements of
the country [12]. Generally, kraft paper is recycled from the recovered and imported
wastage. Only about 30% of the paper in India is recovered due to poor collection
mechanism [12]. According to Indian Paper Mills Association (IPMA), the paper
manufactured from raw wood is available at USD 40 per ton, whereas it costs around
USD 110 in India [12]. All these factors collectively lead to the use of inferior quality
of kraft paper in India. This paper loses its strength even more after usage, and so
good quality of imported kraft paper is mixed with recovered paper during recycling
to strengthen the fibres.

Corrugation

Corrugation is the process of shaping into groves. The corrugated sheet is made by
pasting a corrugated paper over the plain paper. This is attained by virtue of heat and
pressure during the corrugation process, which softens the thermoplastic fibres of the
paper to be corrugated. The basic details of corrugation are depicted in Fig. 4. The
corrugating machine can have an electric or fuel-based heater or steam to heat the
rolls. The corrugated sheet has increased strength due to the double layer of paper

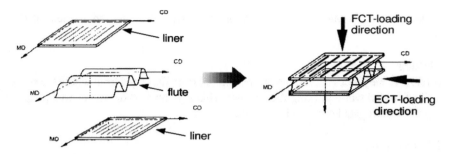

Fig. 4 Understanding basic corrugation

as well as flexibility due to corrugation on one side. It is used for packing irregular shapes during transportation. Corrugated boxes are made by moulding the sheet made by pasting a printed or non-printed sheet over the corrugated sheet/s. The boxes are made by corrugation instead of heavy paper board because the corrugated sheet provides the compressive strength over the paper board of same weight. This helps when boxes are piled over one another. The paper board made box may buckle by its own weight while the corrugated box has the strength to withstand the compressive load due to semicircular pipe like corrugations known as flute in the vertical direction.

Cutting
The corrugated paper or the kraft paper for printing is cut on the cutting machine, which has provision for varying the length of the sheet to be cut.

Printing
There are various methods of printing based on the requirements. For single colour printing, one method is screen printing, which is a manual process and done by making a frame with a tightened silk cloth. Another method for single colour or two colour printing is flexography, in which stereo is made which dips in ink and prints the impression on paper. These two can be done within the premises on the kraft paper similar to the one used for corrugation. While for four-colour printing, offset printing is required. The offset printing requires the thick paper board, and this is done by the third party.

Pasting
The corrugated sheets and the printed sheet are pasted using the gum. After pasting, two-ply sheets are made, and now the rest of the process is to mould the sheet into the box, or any other required shape.

Creasing
Creasing is done to ensure that the sheet folds exactly where it is supposed to fold. Every sheet has to go through creasing twice—horizontal and vertical. Creasing operation also has the cutting edge so that, the extra paper kept to overcome the

non-alignment during pasting is made of the exact size as the requirement of the box.

Slotting
Slotting is the process of removing the material from the edges from where the sheet will be folded while packing the product. This is done to avoid the overlapping of the thick corrugated sheet while folding.

Die-Punching
This process is an alternative to the creasing and slotting operations. Die for a particular size of the box is designed and the pasted sheet is punched on that die. This single operation does horizontal and vertical creasing along with slotting. Also, manufacturing of boxes other than cuboids shape can be done by die-punching. Since this process requires the die for a particular shape and size of the box, the cost of die makes it difficult for small order size box manufacturing. Therefore, die-punching cannot completely replace the creasing and slotting operations.

Stitching/Glueing
The box is stitched using stitching wire. This operation is similar to the stapling operation. Sometimes the metallic stitches may damage the product inside due to sharp edges. To avoid this damage, the box is glued instead of stitching. This glueing process requires another bonding agent.

Strapping
The boxes are piled and bounded together by straps for transportation.

The process of corrugated box manufacturing in the identified industry is shown in Fig. 5.

1.1 Raw Materials of the Industry

1. The major raw material of corrugation plant is bleached or non-bleached recycled kraft paper and paper board. The quality of the paper is specified by two factors: Bursting Factor (BF) or Bursting Strength and area density (GSM). Bursting factor signifies the load at which paper bursts, whereas GSM or grams per square metre shows the weight density of paper, which is also considered while finding the weight of the box. Higher BF and GSM resembles a higher strength of the kraft paper. Kraft paper is used in corrugation as well as for single colour printing, while the duplex paper board is used for four-colour printing.
2. Gum is used for various processes like corrugation, pasting and glueing. During corrugation, the gum is used to paste the corrugated layer over the plain sheet of paper. In pasting operation, multiple corrugated sheets or printed papers are pasted by using the gum.

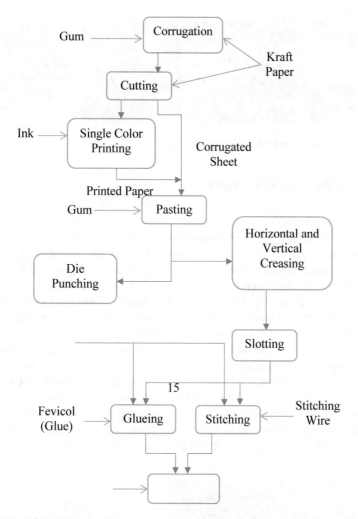

Fig. 5 Material flow diagram of a basic corrugation industry

3. Stitching wire is used to stitch the boxes in the stitching operation. Since this is metallic wire with pointed edges, it sometimes hinders the product inside and may cause corrosion in some cases. Stitching wire is replaced by glue to avoid such losses.
4. Ink is used for single colour printing of front sheet. Ink dip rubber printing block is pressed over the paper, and an impression gets printed on the paper. Single colour printing is done within the facility, while four-colour printed sheets are outsourced.

Figure 6 shows the basic raw materials of the identified industry.

Fig. 6 Kraft paper, gum powder, stitching wire, ink

1.2   Products of the Industry

The industry supplies corrugated sheets, corrugated flaps and corrugated boxes. Corrugated sheets are the paper sheets with one corrugated layer of paper pasted over the other plain layer of paper. Corrugated sheets are more flexible than the plain paper sheet, which is beneficial for wrapping irregularly shaped objects for transportation purposes. Figure 7 shows the products of the industry.

Corrugated flaps are made by punching the sheets and are used to pack the tins from top and bottom. These flaps are used to pack the tins of edible oil from the top.

Corrugated boxes or cartons are packaging material for many products as they facilitate safe handling for the delicate and/or expensive products. A corrugated box is designed on the basis of the material to be packed. In case of water bottles packing, the cost of water bottles is very less and so spending much on corrugated box becomes a burden and so low-quality packing is done. On the other hand, for edible oil and beverages, safe handling is equally important and so box with good strength is designed. Based on these considerations, boxes are designed for various carrying capacities and sizes with various layers, as shown in Fig. 8.

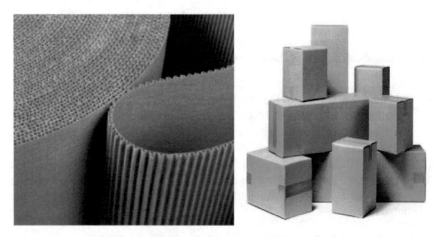

Fig. 7 Corrugated sheet and boxes (personal product image)

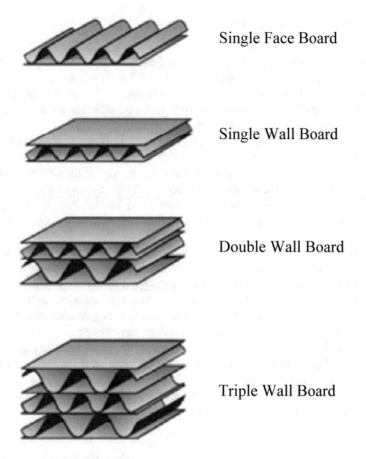

Single Face Board

Single Wall Board

Double Wall Board

Triple Wall Board

Fig. 8 Layers of the corrugated sheet

1. Single face board (SFB): Single faced board or sheet is made of two layers of paper one corrugated and one plain. This is also known as the corrugated sheet.
2. Single wall board (SWB): Single wall board is made of three layers of paper; the inner layer is corrugated, which is covered by a plain sheet on both sides. This sheet can be formed into boxes and generally used for cheaper products with strength in themselves as this sheet can bend easily.
3. Double wall board (DWB): Similar to single wall board, it contains five layers of paper of which two layers are corrugated, and three are plain. The boxes made of this sheet are good in strength.
4. Triple wall board (TWB): This board contains three layers of corrugated sheets and four layers of plain sheets. This sheet is generally used to protect the products from outside impacts. Expensive alcohols are packaged in boxes made of such sheets.

1.3   Types of Flutes of Corrugation

The size of grooves in the corrugation is identified by the flute of corrugation. Bigger the flute size, thicker will be the box. The size of the flute is decided on the basis of the size of the box and the mass of the substance to be filled in the box. The flutes are differentiated on the basis of take-up factor, which is defined as the ratio of non-fluted length to the fluted length of paper [21].

The commonly used flutes of corrugation are A, B, C, E, and F, as shown in Fig. 9.

A-flute is the biggest flute and was developed first of all. It has high cushioning properties for packaging of fragile products. It provides high stiffness and short column crushing strength, which helps in piling of packed cartons one over other.

C-flute is smaller than A-flute. This flute provides crushing resistance, good stacking strength as well as printability. This flute is used in about 80% corrugation industry.

B-flute is smaller than C-flute, but it was developed before C-Flute. As the flute size is small, it gives additional strength to liner during printing. This flute is preferred by automatic corrugation plants where the corrugated sheets are pasted.

E-flute is smaller than B-flute and provides good printability as well as reduced thickness which helps in less storage space. This flute is generally used to replace the paper board boxes due to less material and greater strength.

F-flute is the smallest flute and is made by low-grade paper to avoid solid wastage into landfill. This flute is used when there are no loads on the small boxes like in jewellers shops, display purposes, etc.

Combinations of flutes are used to replace the standard flutes in order to provide strength as well as ease of printing in automated plants [16].

Fig. 9 Types of flutes in the paper corrugation [17]

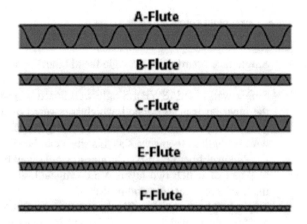

2 Methodology

2.1 Environmental Assessment

The environmental impacts of industries mainly focus on carbon emissions. In this study, Ecological Footprint (EF) is used to assess the environmental impact of an industry which not only imbibes the carbon emissions but also resembles the actual impact of the industry on the planet. Ecological footprint measures the bio-productive land required to meet the needs of resources and energy. The basic formula of ecological footprint is shown in Eq. 2 [8]

$$EF = \frac{Consumption\ Yield}{*} Conversion\ factor \qquad (2)$$

The conversion factor of different types of bio-productive lands is provided in Table 1 [9]. The EF of an industry can be calculated by assessing all the products manufactured in the industry. To calculate the EF of a product, the factors considered were raw materials, transportation of resources, processing energy, manpower, water use and constructed land, etc. The EF of the product is assessed by Eq. 3 [10]:

$$EF_{PRODUCT} = EF_{RM} + EF_E + EF_M + EF_{WM} + EF_T + EF_W + ELF_L. \qquad (3)$$

2.1.1 Ecological Footprint of Raw Material (EF_{RM})

The ecological footprint of raw material is calculated by the embodied energy content of the raw materials. As all the raw materials are processed in some other industry before use in the identified industry, they are not a direct product of the land. Calculating the productive land area for the manufacturing of these raw materials is not in the scope of this study. All the raw materials' ecological footprint was calculated based on either the embodied energy or embodied carbon dioxide of the raw material. The ecological footprint calculation of the raw material is done according to Eq. 4 [11]:

Table 1 Conversion factor (γ) of different types of bio-productive land [9]

Type of land	Conversion factor (γ) (gha/ha)
Cropland	2.52
Built-up land	2.52
Forest land	1.28
CO_2 absorption land	1.28
Pastureland	0.43
Sea productive/Marine land	0.35

$$EF_{RM} = \left(\sum W_j * (1 + \omega_j) * E_{CO2j} \right) * \frac{(1 - L_{oc})}{L_f} * \gamma_{CO2} \qquad (4)$$

where W_j is the weight of jth raw material utilized (tons), ω_j is the waste factor of jth raw material, E_{CO2j} is the embodied carbon dioxide of the jth raw material, i.e. the CO_2 emitted while bringing the raw material in the required state of usage (t_{CO2j}/t_j), L_{oc} is the fraction of CO_2 absorbed by the ocean, L_f is the capacity of forest land to absorb CO_2 (t_{CO2}/ha), γ_{CO2}(=1.28) is the conversion factor of CO_2 absorption land (gha/ha).

When embodied energy of the substance is given, the embodied carbon dioxide is calculated by Eq. 5.

$$E_{CO2j} = E_{Ej} * ¥ \qquad (5)$$

where E_{Ej} is the embodied energy of the jth material (MJ/ton), and ¥ is the factor of conversion of embodied energy to embodied carbon dioxide based on the total primary energy use and emissions of the country (t_{CO2}/MJ_p).

As the wastage is only of kraft paper and that too is recycled, therefore the impact of waste material is just due to the transportation of waste material for recycling, which is discussed in Sect. 2.1.4.

2.1.2 Ecological Footprint of Energy (EF$_E$)

The ecological footprint of energy is calculated based on the source of energy. When the source of energy is some kind of fuel, then the carbon dioxide generation of the fuel is considered, and the land required to mitigate the emitted CO_2 is calculated to obtain the ecological footprint. When the source of energy is a high grade of energy like electricity, the primary emissions during energy generation of the region are considered to calculate the actual CO_2 emissions using Eq. 6.

$$EF_E = \left(\sum E_j * E_{CO2j} \right) * \frac{(1 - L_{oc})}{L_f} * \gamma_{CO2} \qquad (6)$$

where E_j is the amount of direct energy of jth type used in industry (MJ or MWh), E_{CO2j} is the CO_2 emitted due to energy usage of jth source per unit (t_{CO2}/MJ or t_{CO2}/MWh).

2.1.3 Ecological Footprint of Manpower

There is no consistency when manpower is looked upon as a resource. Someday some labour may be on leave while on the other day the production might be less. Hence, to calculate the impact of manpower, the number of labours and workdays together cannot describe the actual scenario. Therefore, the number of labour-hours is taken

Table 2 Per capita raw material for food preparations in India

Food goods	Annual consumption	Annual EF (gha)
Cereals	111.36 kg	0.117
Pulses	10.8 kg	0.039
Vegetables	100.8 kg	0.157
Beef	0.72 kg	0.011
Mutton	0.96 kg	0.006
Milk	64.8 L	0.062
Fish	3.024 kg	0.030
Fruits	7.848 kg	0.047
Edible oil	10.2 L	0.023
Wood	51.6 kg	0.028
LPG	22.8 kg	0.025
Kerosene	4.8 L	0.004

into account, which can be obtained from the attendance sheet of the labours. To estimate the ecological footprint of the labour-hour, the metabolic energy requirements of the labour during working are taken into account along with the total energy intake using Eq. 7. It is estimated that during work hours, 155.3 kcal/h of energy is required [18]. Also, the daily energy intake of an individual in India is 2400 kcal/day [5]. The food consumption and their ecological footprints are calculated by [10]. These are showcased in Table 2.

The ecological footprint of labour-hour is calculated as

$$\text{EF}_M = \frac{E_f}{E_d} * \frac{1}{365} * \left\{ \sum \left(\frac{C_i}{Y_i} * \gamma_i \right) + \sum \left(C_{fj} * \lambda_{fj} \right) * \frac{(1 - L_{oc})}{L_f} * \gamma_{CO2} \right\} \quad (7)$$

where E_h is the energy required during one hour of work (kcal), E_d is the energy consumed in a day (kcal), C_i is the amount of ith food material consumed by a person in one year (kg/year), Y_i is the yield of ith food material (kg/ha), γ_i is cropland conversion factor (=2.32), C_{fj} is the consumption of jth fuel for cooking (kg/year or L/year), λ_{fj} is the emission of jth fuel (kgCO$_2$/kg or kgCO$_2$/L).

2.1.4 Ecological Footprint of Transportation (e_T)

Transportation involves movement of raw material from the supplying industry to the identified industry, paper waste transportation from the industry to recycling plant and transportation of manpower. Generally, raw materials and waste paper are transported by heavy-duty trucks. The ecological footprint of material transport can be calculated by Eq. 8 [10]:

$$\mathrm{EF}_T = \left(\sum \frac{W_j * D_j}{T_{\mathrm{HDT}}} + \frac{W_i * D_i}{T_{\mathrm{HDT}}} \right) * \eta_{\mathrm{HDT}} * \lambda_f \frac{(1 - L_{oc})}{L_f} * \gamma_{\mathrm{CO2}} \qquad (8)$$

where W_j is the weight of jth raw material transported (ton), D_j is the average distance of transportation of jth raw material (km), W_i is the weight of ith waste material transported for recycling (ton), D_i is the average distance of transportation of ith waste material (km), T_{HDT} (11 ton) is capacity of heavy-duty truck, η_{HDT} (0.222 kg/km) is fuel economy [2].

Half of the labour lives within the premises of industry, while the other half would travel from nearby areas. Thus, the travel impact of labour is of negligible significance in this study.

2.1.5 Ecological Footprint of Water

Water is supplied by Maharashtra Industrial Development Corporation (MIDC), Khamgaon. It is consumed for making gum and other human needs. The usage of water is considered for processing only. The ecological footprint of water can be calculated using Eq. 9 [4]:

$$\mathrm{EF}_{\mathrm{w}} = C_{\mathrm{w}} \cdot \left\{ E_{\mathrm{w}} \cdot \alpha_e \cdot \frac{(1 - A_{oc})}{A_f} \right\} e_{\mathrm{CO_2 land}} \qquad (9)$$

where C_{w} is total water use during all phases of RSPV system (m^3); E_{w} is electricity consumption (kWh/m^3); α_e is emission factor of electricity.

2.1.6 Ecological Footprint of Land (E_L)

The direct land used by the industry for carrying out manufacturing processes is the built-up land. The ecological footprint of direct land is calculated for the period of 60 years considering the life of the industry, according to Eq. 10.

$$\mathrm{EF}_L = \frac{A}{60 * P_{\mathrm{A}}} * \gamma_{\mathrm{BL}} \qquad (9)$$

where EF_L is the ecological footprint per ton of product (gha/ton), A is the area of built-up land (ha), P_{A} is average annual production output of all products (ton), γ_{BL} is the conversion factor of built-up land.

2.2   Economic Assessment

The economic assessment is equally important to verify the viability of suggested ecological footprint reduction measures. The economic assessment of the cost of production of the products per-unit mass is done based on the factors: raw material, energy, labour, repair and maintenance, transportation and water.

2.2.1   Cost of Raw Materials

The average cost of raw material is calculated based on the year-long variation of prices. The average cost of kraft paper includes costs of kraft paper for corrugation and single colour printing and duplex board for four-colour printing as all these papers are of same bursting factor. Similarly, the average cost of modified starch includes starch for corrugation and starch for pasting. All other raw materials are of the same kind, and their average cost is calculated based on the year-long variation.

2.2.2   Cost of Energy

All the energy requirements of the industry are currently met by electricity supplied by grid alone. The cost of electricity was averaged based on energy charges, monthly rentals, time of day tariff, fuel adjustment charges, power factor penalties/incentives, electricity duty, wheeling charges, tax on sale and other charges.

Energy charges are the per-unit charge of electricity charged according to the category specified by Maharashtra State Electricity Distribution Company Limited (MSEDCL) as LT-V B II which is Low Tension Industrial Purposes for general usage over 20 kW supply. Monthly rentals include the fixed monthly charges for using the electricity provided by the company. Time of day tariff (t.o.d tariff) includes the charges and incentives on time-bound usage to improve the uniformity in demand along the day. During peak hours the t.o.d is charged while during low loads incentives are given on energy usage. Power Factor (PF) incentives are provided to encourage industries to keep the Power Factor close to 1, while they are charged penalties for poor PF. Wheeling charges are the transmission charges levied on every unit consumed. All these charges are considered to calculate not only the actual rate of electricity from the grid but also when replaced by some other source of energy.

2.2.3   Cost of Labour

Cost of manpower or labour is averaged based on the type of labour and hours of labour required. This cost is calculated for labour-hours based on the wages of skilled and unskilled labours and the number of labours.

2.2.4   Cost of Transportation

The various raw materials being transported from various sources in variable amounts are considered to calculate the average cost of transportation of raw materials. Cost of transportation of waste kraft paper to the recycling plant is also considered.

2.2.5   Cost of Water

Water needs for the industrial purpose are less, and so the impact is less as well. Water is provided to the industry by MIDC. The cost of water consumed is averaged based on the monthly rentals for using the service and the cost of the number of units of water consumed.

2.2.6   Pay Back Period

Payback period of the suggested modes is calculated based on the variable tariffs. The rental tariffs will, however, remain as long as the connection of electricity from grid exists in case of alternative energy modes. Payback period in raw material saving mode will be directly calculated based on the cost of machinery.

2.3   Suggested Modes for Ecological Footprint Reduction

2.3.1   Replacing Existing Grid Electricity Consumption Completely with Solar PV Grid-Connected Electricity

As the total energy needs of the industry are met by electricity from the grid alone, the replacement of source of electricity with the interconnected solar PV-based electricity and the grid electricity through inverter would make the industry self-sufficient for its energy requirements. Figure 10 shows the basic pictorial representation of Mode 1.

The analysis of this mode was done by using RETScreen, which has the database of the temperature variation, wind speed and other regional parameters sourced from NASA as shown in Fig. 11.

2.3.2   Replacing Existing Grid Electricity Consumption for Corrugation Heating with LPG

About 60% of the total energy need of the industry is required to provide heat to the corrugation roll. There are three electric corrugation heaters of 12 kW each which accounts for 36 kW while the total capacity of the plant is 55 kW. Electricity is

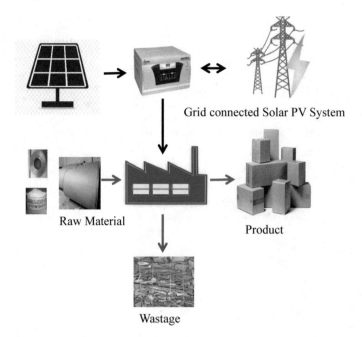

Grid connected Solar PV System

Raw Material

Product

Wastage

Fig. 10 Pictorial representation of Mode 1

generated in India with the overall efficiency of 32%, which means it takes about three equivalent units of primary energy to generate one unit of electricity. For corrugation, this electricity is again converted to heat; therefore, it is better to generate the heat at the required place by burning the fuel. Many medium-scale corrugation industries run boiler to meet the needs of heat for corrugation. But, the boiler is preferred when the heat is required continuously for long hours. LPG is easily available in India for industrial usage. Thus, replacing the part of electricity which is meant for heating by LPG is suggested as shown in Fig. 12, which would require replacing the electric heaters with gas burners of equivalent size. This modification won't require heavy investments, but may provide both economic and environmental benefits.

2.3.3 Replacing the Grid Electricity with Solar PV Grid-Connected Electricity and LPG for Heating Purposes

In this mode, the grid-connected electricity is completely replaced by LPG for heating the corrugation roll, and solar PV grid-connected electricity to meet other needs of electricity as shown in Fig. 13. As this mode has two modifications, implementation of this mode can also be done in parts to ease the investments.

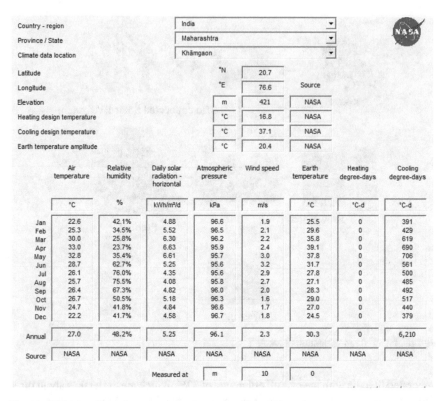

Fig. 11 RETScreen-based weather data of the selected location

2.3.4 Changing the Flute Size of the Corrugated Sheet

All the earlier modes are meant to minimize the energy impact on the environment, while this mode suggests the change of size of a flute to minimize the requirement of paper and meet the requirements of strength as well. The flute size used by industry for corrugation is B-Flute. The double-layered box requires five layers of kraft paper. If the flute size is increased from B-flute to A-flute, the paper required will be of higher strength so that the flutes do not shrink. The analysis of two corrugated sheets with A and B flutes is done based on the Bursting Strength (BS) of the corrugated sheet based on IS 1060-1 standards as shown in Eq. 11.

$$BS = BF*GSM \tag{10}$$

where BF—Bursting Factor, BS—Bursting Strength (g/cm^2) and GSM—grams per square metre (g/m^2).

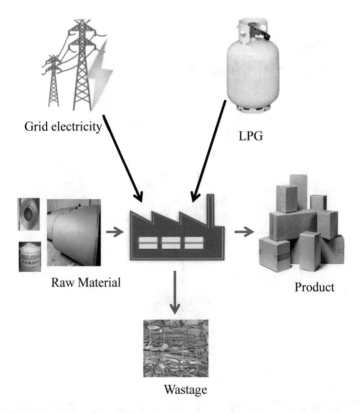

Fig. 12 Pictorial representation of Mode 2

3 Industry Survey

The industry identified is M/S Vishal Packaging, manufacturer of corrugated sheets and boxes which is located in Khamgaon, Maharashtra, India. The location of the industry is shown in Fig. 14. The basic profile of the industry is given in Table 3. The per-unit basic cost of these resources are shown in Table 4. As the rates may vary owing to many factors, these rates are calculated for the financial year 2018–19. Also, the rate of transport cannot be generalized as they are dependent on the variation of vendors for different quantity of raw materials being supplied.

During the survey, the main raw material of the industry was identified as kraft paper. The kraft paper of 16 BF is generally used for maximum purposes in the industry. This kraft paper is of inferior quality which means the recycled paper used by the industry contains weak fibres which signify that the fibres are recycled more and more.

Grid connected solar PV

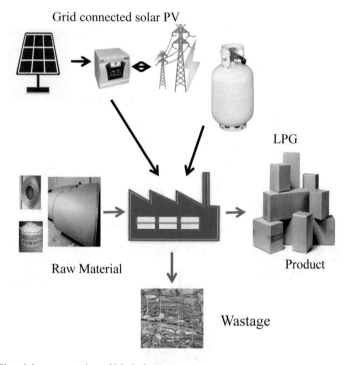

LPG

Raw Material

Product

Wastage

Fig. 13 Pictorial representation of Mode 3

Fig. 14 Google Maps Image of the selected industry

Table 3 Basic details of the selected industry

Name	Vishal packaging
Type of job	Manufacturing of corrugated sheets and boxes
Location	D-25, MIDC, Khamgaon (M.S)
Types of raw materials	Paper, gum, stitching wire, glue, ink, water
Type of energy required	Electricity for driving motors and heating
Number of employees	12
Total energy input	55 MWh
Total paper rolled	480 tons

Table 4 Cost of raw materials

Sr No	Inputs	Unit	Rate	Transportation cost
1	Kraft paper and paper board	Rs./kg	24.85	0.9
2	Modified starch	Rs./kg	63.27	1.75
3	Stitching wire	Rs./kg	36.41	2.96
4	Fevicol	Rs./kg	41.42	0
5	Electricity	Rs./kWh	7.36	0
6	Labour	Rs./Labour hour	33	0
7	Water	Rs./m^3	30.23	0

*Costs of raw material and transport are subjected to market change
*Costs of electricity and water are inclusive of taxes and rentals

4 Data Analysis

4.1 Ecological Footprint

The ecological footprint of resources and energy sources is calculated as tabulated in Table 5.

The ecological footprint of all the required resources is calculated based on the embodied energy of these resources. The embodied energy is used to convert and calculate the equivalent GHG emissions based on the sources of primary energy used by the country.

Ecological footprint of various products (per ton) of the industry based on the amount of energy and resources currently consumed is shown in Table 6 grouped on the basis of raw material, energy, manpower, transportation and built-up land requirements. The product-wise ecological footprint is calculated, as shown in Table 7.

a. **Sample Calculation of Ecological Footprint**

Table 5 EF of input energy and resources

Resource	Unit	EF of resources	EF of transportation
Kraft paper	gha/ton	0.4280	0.0031
Modified starch	gha/ton	1.1972	0.0042
Stitching wire	gha/ton	1.095	0.0099
Glue	gha/ton	0.6836	0.0001
Grid electricity	gha/kWh	0.0004	–
PV electricity	gha/kWh	0.00002	–
LPG	gha/ton	1.1	–
Water	gha/m^3	0.394	–
Land	gha/ton of product	0.00000588	

Table 6 Resources required per ton of product

Resource	SFB	SWB	DWB	TWB
Kraft paper (kg/ton)	981.65	958.85	965.20	971.18
Starch (kg/ton)	18.34	25.10	26.33	26.71
Stitching wire (kg/ton)	0	16.04	7.12	2.10
Fevicol (kg/ton)	0	0	1.34	0
Total weight	1000	1000	1000	1000
Grid electricity (kWh/ton)	146.99	159.28	149.07	149.24
Manpower (labour-h/ton)	99.15	107.44	100.54	100.67
Water (L/ton)	73.39	100.39	105.33	106.85

Table 7 Ecological footprint gha/ton product

	SFB	SWB	DWB	TWB
EF (raw material)	0.44226	0.45818	0.45352	0.45013
EF (energy)	0.05879	0.06371	0.05962	0.05969
EF (manpower)	0.01117	0.01211	0.01133	0.01134
EF (transportation)	0.00313	0.00324	0.00318	0.00315
EF (land)	0.0000058	0.0000058	0.0000058	0.0000058
Ecological Footprint	**0.51537**	**0.53725**	**0.52767**	**0.52433**

The sample calculation of EF of the product is explained below:
Consider double wall board-based corrugated box,

i. Ecological Footprint of Raw Material

Kraft Paper:

Embodied GHG of carbon dioxide equivalent $= 1.29$ t_{CO2}/t kraft paper
The capacity of the land to absorb $CO_2 = 2.7$ tCO_2/t
Conversion Factor, $\gamma_{CO2} = 1.28$
The proportion of CO_2 absorbed by ocean $= 0.3$
EF of Kraft Paper $= \frac{1.29}{2.7} * (1 - 0.3) * 1.28$
EF of Kraft Paper $= 0.4281$

Similarly, based on the embodied energy and/or embodied CO_2, EF of all the raw materials can be calculated using Eq. 3 and Eq. 4.

ii. Ecological Footprint of Energy

Grid Electricity:
Based on sources like coal, nuclear, hydro, etc., the average emissions per-unit electricity generation is 0.001186 tCO_2/kWh.
   Therefore,

$$EF \text{ of Grid electricity} = \frac{0.001186}{2.7} * (1 - 0.3) * 1.28$$
$$EF \text{ of Grid electricity} = 0.0004 gha/kWh$$

Similarly, EF for other energy resources can be calculated using Eq. 5 and as tabulated in Table 5.

iii. Ecological Footprint of transportation

$$EF \text{ of Transportation} = \frac{0.222 * 3.17}{11 * 2700} * (1 - 0.3) * 1.28$$
$$EF \text{ of Transportation} = 2.12 \times 10^{-5} gha/ton \text{ - } km$$

Based on the average distance of transportation, EF (per ton) can be calculated.

iv. Ecological Footprint of Land

The built-up land is considered to serve for 60 years. Based on average annual turnover (in tons of products), the impact of land on the product is calculated (refer Sect. 0)

$$Land = 0.07 ha$$
$$Turnover = 500 ton/year$$

Table 8 Sample EF calculation

Resources required	Weight (kg)	EF (gha/ton)	EF Transportation (gha/ton)	EF (gha)
Kraft paper (kg/ton)	965.20	0.4280	0.00311	0.41619
Starch (kg/ton)	26.33	1.19722	0.00424	0.03163
Stitching wire (kg/ton)	7.12	1.09511	0.00988	0.00787
Fevicol (kg/ton)	1.34	0.68361	0.00011	0.00091
Total weight	1000	0	0	0
		(gha/unit)		
Grid electricity (kWh/ton)	149.06	0.0004	0	0.05962
Manpower (h/ton)	100.55	0.00011	0	0.01133
Land	0.0000058	0.0000058	0	0.0000058
Water (L/ton)	0.11	0.8533	0	0.000089
Total EF (gha/ton of product)				0.52767

$$\text{EF of Land} = \frac{0.07}{60 * 500} * 2.52 \text{gha/ton}$$

For ecological footprint of manpower, refer Sect. 0 in methodology.

The ecological footprint of DWB box is calculated on the basis of raw materials required and their individual ecological footprints (Table 8)

b. **Sample Calculation of Industrial Sustainability Index (ISI)**

ISI is calculated based on Eq. 1

$$\text{ISI} = \frac{\text{RVA} * \text{EMP}}{\text{CO}_2 \text{emissions}}$$

$$\text{RVA} = 4.912 \text{Million Rupees/year} \ldots \text{(data from industry for FY 2018}-19)$$

$$\text{EMP} = 13$$

$$\text{CO}_2\text{emissions} = \text{Energy Consumed through various sources} * \text{Emission factor}$$

$$\text{CO}_2\text{emissions} = 55, 150 \text{ kWh} * 0.001186 \text{tCO}_2/\text{kWh} = 65.4 \text{tCO}_2$$

$$\text{ISI} = \frac{4.02782 \text{ Million Rupees} * 13 \text{persons}}{65.4 \text{tCO}_2} \quad \text{ISI} = 0.80 \text{ units}$$

The environmental and economic analysis of the industry is done for the existing technology and suggested modes. The effects of the earlier discussed modes in Sect. 1.2 are explained as follows:

c. **Mode 1 (Replacement of grid-based electricity with grid-connected solar PV electricity)**

Table 9 Basic details of solar PV cells (Mode 1)

Type of solar PV	Poly-SI
Model (manufacturer)	Poly-SI-TSM-PC05 (Trina Solar)
Efficiency	14.4%
Nominal operating temperature	45 °C
Power capacity/panel	235 W
Power capacity	32.9 kW
Number of panels	140
Total electricity exported	55.205 MWh
Solar collector area	233 m^2

In this mode, all the energy needs of the industry are suggested to be met by solar PV-based electricity connected by grid instead of grid-supplied electricity.

The analysis of this mode is done using RETScreen software, which allows the user to estimate the cost to meet the energy requirements by various renewable sources of energy and also estimates the carbon savings against the conventional grid-supplied electricity. The software also enables the user to calculate the payback period of the investments made.

The details of the solar panels to be installed to meet the complete needs of energy are shown in Table 9. To generate about 55 MWh in one year, 32.9 kW of poly-Si solar panels will be required based on the needs of industry and solar radiation of the region. The analysis suggests that 140 units of 235 W solar panels with 14.1% efficiency will be required to fulfil the energy requirements. The total area required to set up solar panels is around 233 m^2. Figure 15 shows the monthly electricity generation at the location of industry on the basis of weather conditions of that location around the year.

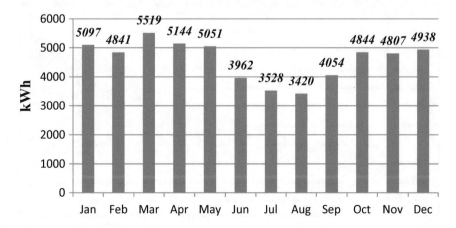

Fig. 15 Monthwise electricity export to grid

Table 10 Economic analysis (Mode 1)

Total investment	Rs. 2,138,500
Electricity export rate	Rs. 7390/MWh
Yearly savings from electricity export	Rs. 407,963
Payback period	5.3 years

Table 11 Net savings of money and GHG

IRR	18.3%
Project life	20 years
Net Present Value (NPV)	Rs. 6,106,067
Benefit-Cost (B-C) ratio	3.86
GHG savings per year	65 tCO$_2$
GHG savings over the life of the project	1309 tCO$_2$

The cost economy of this mode can be easily depicted in Table 10. The total investment cost of the solar panels is expected to be around Rs. 2,138,500. The cost of each unit is averaged at Rs. 7390/MWh. The annual maintenance cost, which includes the cleaning of the panels, is projected at Rs. 3000. The end of project life is assumed to be Rs. 100,000. The total electricity income to be generated by the export of electricity can be predicted to be around Rs. 407,963/year.

Table 11 explains the net savings of money and GHG around the year. At the IRR of 18.3%, the net savings along the year is calculated to be Rs. 305,303/year. The net GHG reduction in one year is about 65 tCO$_2$ equivalent. Considering the life of the project of about 20 years, the total GHG savings would be around 1309 tCO$_2$. It also suggests that the simple payback period of this investment is around 5.3 years, as shown in Fig. 16.

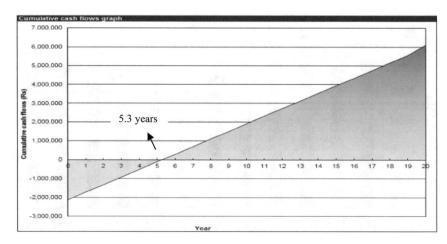

Fig. 16 Payback period (Mode 1)

d. Mode 2 (Replacement of Electric heater for corrugation with LPG burner)

Corrugation heaters are used to heat the corrugation roll to make plastic deformation of the fluted paper so that the paper must retain the shape of corrugation. Three electric heaters of 12 kW each are used to carry out the corrugation process. These electric heaters contribute to about 60% of the total energy requirements of the industry. As the main source of electricity generation in India is coal [6], the overall efficiency of energy utilization is approximately one-third. Therefore to reduce the emissions for heating, use of LPG fuel is suggested which will be cost-effective as well as it will reduce emissions. By replacing the grid-based electric heaters with LPG burners, the total GHG savings are expected to be $30.70tCO_2$/year, as shown in Table 12.

While calculating the cost savings, it is essential to consider that the carbon layer will deposit on the burner during burning, this will have to be removed from time to time. This, therefore, requires maintenance periodically whose cost needs to be incorporated. The total cost savings due to the use of LPG may stand up to Rs. 52,153.2 as depicted in Table 13.

e. Mode 3 (Replacing electric heater with LPG gas burner and grid electricity with grid-connected solar PV electricity to meet other energy needs)

As the investment cost of replacing all energy needs of the industry by solar PV is quite high, the cost-effective mode which replaces the heating requirements with LPG gas and other electrical needs by grid-connected solar PV is suggested. This analysis is done using RETScreen for solar PV analysis, and the LPG analysis is done

Table 12 Total GHG savings of Mode 2

	Electricity	Unit	LPG	Unit
GHG emissions	0.001186	tCO_2/kWh	0.0000718	tCO2/MJ
Energy requirements	33,100	kWh/year	119,160	MJ/year
Total GHG emissions	39.26	tCO2/year	8.56	tCO2/year
Total GHG savings	30.70			tCO2/year

Table 13 Net savings of Mode 2

	Electricity	Unit	LPG	Unit
Energy required for heating	33,100	kWh/year	119,160	MJ/year
The unit cost of energy	7.39	Rs/kWh	1.48	Rs/MJ
Cost of energy	244,609	Rs/year	176,855.8	Rs/year
Cost savings	67,753.20			Rs/year
Maintenance			300	Rs/week
Total maintenance cost			15,600	Rs/year
Net savings	52,153.0			Rs/year

differently. To calculate the payback period, both the improvements are collectively and differently stated.

i. Solar PV Calculations

Similar to the analysis of Mode 1, this analysis is done using RETScreen as discussed below.

To generate 22 MWh of electricity in one year, poly-Si-based solar PV is suggested. Fifty-six solar PV panels of 235 W each will be required. The power rating of the panels together is 13.16 kW, with 14.1% efficiency as specified in Table 14.

The total investment of solar PV panels is around Rs. 855,400. The annual electricity export income is expected to be Rs. 163,185, and the annual GHG savings is around 26 tCO_2 equivalent. With the total life expectancy of 20 years, total GHG savings are expected to be 524 tCO_2 equivalents, as shown in Table 15. Similar to Mode 1, with 18.3% of IRR, the simple payback period is expected to be 5.3 years for the solar PV-based energy input.

ii. LPG Calculations

The LPG-based results are similar to the Mode 2 results. Individually, LPG-based setup with the capital investment of approximately Rs. 60,000 gives annual savings of about Rs. 53,153.2/year. This gives the simple payback period of 1.15 years. The annual GHG savings by LPG are estimated to be around 30.7 tCO_2/year.

Table 14 Basic details of solar PV cells (Mode 3)

Type of solar PV	Poly-SI
Model (manufacturer)	Poly-SI-TSM-PC05 (Trina Solar)
Efficiency (%)	14.4
Nominal operating temperature (°C)	45
Power capacity/panel (W)	235
Power capacity (kW)	13.16
Number of panels	56
Total electricity exported (MWh)	22.082
Solar collector area (m^2)	93

Table 15 Economic analysis of solar PV (Mode 3)

Total investment	Rs. 855,400
Electricity export rate	Rs. 7390/MWh
Yearly savings from electricity export	Rs. 163,185
Payback period	5.3 years
GHG savings per year	26 tCO_2
GHG savings over the life of the project	524 tCO_2

Table 16 Cumulative investments and savings of Mode 3

	Solar PV	LPG	Total
Investment (Rs)	855,400	60,000	915,400
Annual savings (Rs/year)	163,185	52,153.2	215,338.2
Individual payback (years)	5.24	1.15	
Cumulative payback (years)			4.25
Annual GHG savings (tCO$_2$/year)	26.2	30.7	56.9

The cumulative investments and savings of money and GHG along with payback period are summarized in the table below.

Table 16 shows the total investment of solar PV along with LPG-based gas burner setup stands at Rs. 915,400. With the annual savings of Rs. 215,338.2, the cumulative payback period is estimated to be 4.25 years.

5 Results

The analysis of savings of cost and emissions has resulted in a reduction in cost as well as the ecological footprints of all the products in all the modes as discussed below:

f. **Mode 1**

The results of Mode 1 are tabulated in Table 17 based on Tables 9, 10 and 11.

The EF and cost of SFB are least as it does not contain stitching wire or glue, and the process is also small. Thus, energy requirements are also less. This can be easily seen from Table 17. A decreasing trend of EF in wall boards can be seen with

Table 17 Product-based EF and cost per ton for Mode 1

Raw material	SFB	SWB	DWB	TWB
EF (raw material) (gha)	0.44223	0.45813	0.45347	0.45008
EF (energy) (gha)	0.00293	0.00318	0.00298	0.00298
EF (manpower) (gha)	0.01115	0.01208	0.01131	0.01132
EF (transportation) (gha)	0.00313	0.00324	0.00318	0.00315
EF (land) (gha)	0.000006	0.000006	0.000006	0.000006
Total EF (gha)	0.45946	0.47666	0.47096	0.46755
Cost (raw material) (Rs)	25,069.96	25,765.02	25,458.64	25248.02
Cost (energy) (Rs)	273.27	296.11	277.11	277.43
Cost (manpower) (Rs)	3272.03	3545.56	3318.10	3322.01
Cost (transportation) (Rs)	915.59	954.40	935.85	927.04
Total cost (Rs)	29,530.85	30,561.08	29,989.70	29,774.5

Table 18 ISI for Mode 1

RVA	4.3331	Million Rs
GHG	1.1	tCO_2
EMP	13	person
ISI	51.2	Million Rs-persons/tCO_2

the increasing number of layers. It can be majorly because the use of stitching wire per-unit weight of the box decreases significantly.

Also, the ISI of the industry for this mode is calculated as summarized in Table 18. It can be compared with the ISI of the industry with existing conditions from Sect. 0.

g. **Mode 2**

Replacing just the electric heaters with the LPG based gas burners is the cheapest of all modes discussed. This mode just reduces the impact of energy used for heating during corrugation, while the rest of the energy needs will be still met by grid-supplied electricity. Thus, the impact of this mode is least of all the energy modes. The impact of this mode can be depicted in Table 19.

It can be observed that the impact of economic and environmental benefits of this mode is less as compared to Mode 1, due to the use of LPG for heating purposes only. It should also be considered that the investment of this mode is less, and accordingly, the impact can be observed.

The ISI calculation for this mode is summarized in Table 20. The RVA has increased due to savings of energy cost due to the use of LPG over grid-connected electricity. Also, the emissions impact is reduced due to use of LPG directly instead of converting low grade of energy (heat) to high grade of energy (electricity) which has less overall efficiency for heating.

h. **Mode 3**

Table 19 Product-based EF and cost per ton for Mode 2

Raw material	SFB	SWB	DWB	TWB
EF (raw material) (gha)	0.44226	0.45818	0.45352	0.45012
EF (energy) (gha)	0.03071	0.03327	0.03114	0.03118
EF (manpower) (gha)	0.01115	0.01208	0.01131	0.01132
EF (transportation) (gha)	0.00313	0.00324	0.00318	0.00315
EF (land) (gha)	0.000005	0.000005	0.000005	0.000005
Total EF (gha)	0.48726	0.50680	0.49917	0.49580
Cost (raw material) (Rs)	25,069.96	25,765.02	25,458.64	25,248.02
Cost (energy) (Rs)	906.24	982.01	919.01	920.09
Cost (manpower) (Rs)	3272.03	3545.56	3318.10	3322.01
Cost (transportation) (Rs)	906.74	945.70	927.13	918.28
Total cost (Rs)	30,154.99	31,238.28	30,622.86	30,408.39

Table 20 ISI for Mode 2

RVA	4.07998	Million Rs
GHG	34.627	tCO_2
EMP	13	person
ISI	1.5317	Million Rs-person/tCO_2

This mode is a combination of both the above modes, which proposes the use of LPG for corrugation heating and grid-connected solar PV-based electricity for other energy needs. The impact of this mode can be compared by the base mode and can be seen as considerable improvements in ecological impacts. The details of EF and cost are discussed in Table 21.

The product-wise results of this mode can depict the actual impact of using both solar PV and LPG against the use of LPG alone (Mode 2). This makes it clear that the total investment of this mode is less than Mode 1 but undoubtedly, greater than Mode 2. The ISI calculation of the industry based on this mode is calculated in Table 22. The ISI of this mode lies between that of Mode 1 and Mode 2 and is far more than the existing practices.

i. **Mode 4**

Table 21 Product-based EF and cost per ton for Mode 3

	SFB	SWB	DWB	TWB
EF (raw material) (gha)	0.44226	0.45818	0.45352	0.45012
EF (energy) (gha)	0.00874	0.00947	0.00886	0.00887
EF (manpower) (gha)	0.01115	0.01208	0.01131	0.01132
EF (transportation) (gha)	0.00313	0.00324	0.00318	0.003154
EF (land) (gha)	0.000005	0.000005	0.000005	0.000005
EF (gha)	0.46530	0.48300	0.47690	0.47350
Cost (raw material) (Rs)	25,069.96	25,765.02	25,458.64	25,248.02
Cost (energy) (Rs)	573.73	621.69	581.80	582.49
Cost (manpower) (Rs)	3272.03	3545.55	3318.09	3322.01
Cost (transportation) (Rs)	906.74	945.70	927.12	918.28
Total cost (Rs)	29,822.47	30,877.97	30,285.67	30,070.80

Table 22 ISI for Mode 3

RVA	4.3564	Million Rs
GHG	6.13	tCO_2
EMP	13	Person
ISI	9.23	Million Rs-persons/tCO_2

Table 23 Details of B-flute DWB and A-flute SWB

			B-flute	A-flute
Paper		GSM	100	150
		BF	16	24
		BS	1.6	3.6
Corrugated sheet		TF	1.36	1.56
		Liners	3	2
		Flutes	2	1
		GSM	572	534
		BS	–	–

Using the bigger flute requires an improved quality of the paper. For B-flute, the paper used in the industry is 100 GSM 16 BF. For A-flute, paper requirements will be 150 GSM 24 BF. This will improve the Bursting Strength of Paper.

The weight of A-flute SWB is less than B-flute DWB by 6.64%, as mentioned in Table 23. This reduction in raw material will further reduce the other resources required. The details of raw material required for the B-flute and A-flute per ton of the corrugated box are given in Table 24.

5.1 Comparison of Modes

The comparison of all the modes is made in Tables 25 and 26, based on investments, EF and ISI along with cost savings.

The comparison of all the modes clearly shows that the maximum ecological impact with the least capital investment is attained in Mode 2, while maximum EF reduction is expected by adapting Mode 1. The EF reduction of Mode 3 compared to Mode 1 is higher when considered from a capital investment point of view. Also, it should be considered that the ISI improvement in Mode 1 is maximum because it not only reduces the ecological impact but also reduces the cost of energy drastically which increases the revenue and thereby improves ISI.

The percentage share of resources and services analysed for different modes show that the major share of EF comes from raw material in all the modes, while the share of energy is maximum in the existing mode and considerable change can be seen in suggested modes.

The product-wise comparison of all the suggested modes shows that the impact of solar PV-based Mode 1 is highest on EF of energy which is approximately 95% for all the products, while for Mode 2 and Mode 3 it is 47.7% and 85%, respectively. Also, cost savings follows the same trend.

The share of EF of various resources and services for all the modes is shown in Fig. 17.

Table 24 Resources and EF per ton for Mode 4

Raw material	SWB A-flute	DWB B-flute	Required SWB A-flute
Kraft paper (kg/ton)	972.18	965.20	907.60
Starch (kg/ton)	16.96	26.33	15.83
Stitching wire (kg/ton)	10.84	7.12	8.51
Fevicol (kg/ton)	0	1.34	1.62
Total weight (kg)	1000	1000	933.57
Grid electricity (kW-h/ton)	107.67	149.065	100.51
Manpower (h/ton)	72.62	100.55	67.80
Land (ha)	0.00000588	0.00000588	0.00000549
Water (L/ton)	67.86	105.33	63.35
EF (raw material)	0.44843	0.45352	0.41864
EF (energy)	0.04306	0.05962	0.04020
EF (manpower)	0.00818	0.01133	0.00764
EF (transportation)	0.00320	0.00318	0.00299
EF (land)	0.00000588	0.00000588	0.00000549
Ecological footprint (gha)	0.50289	0.52767	0.46948

Table 25 Comparison of all the energy modes

	Existing	Mode 1	Mode 2	Mode 3
EF (energy) (gha)	22	1.1	4.185	3.272371
Total EF (gha)	239.28	218.39	228.77	220.55
Yearly savings (gha)	–	20.91	10.52	18.74
ISI (MillionRs-person/tCO$_2$)	0.80	51.20	1.53	9.23
Investment (Rs)	–	2,138,500	60,000	915,400
Yearly savings (Rs)	–	407,963	52,153.20	215,338.2
Payback (years)	–	5.3	1.15	4.25

6 Conclusions and Scope of Future Work

The paper corrugation-based packaging industry is surveyed. Three energy modes, based on the potential of grid-connected solar PV and LPG, are analysed to reduce

Table 26 Product-based comparison of different modes on EF and cost

		SFB	% change	SWB	% change	DWB	% change	TWB	% change
Existing	EF (energy) (gha/ton)	0.0588	0	0.063	0	0.059	0	0.059	0
	Total EF (gha/ton)	0.5154	0	0.537	0	0.527	0	0.527	0
	Cost (Rs/ton)	30,345	0	31,443	0	30,815	0	30,815	0
Mode 1	EF (energy) (gha/ton)	0.0029	95	0.003	95	0.003	95	0.003	94.9
	Total EF (gha/ton)	0.4595	10.8	0.476	11.27	0.471	10.74	0.467	11.3
	Cost (Rs/ton)	29,530	2.68	30,561	2.80	29,989	2.68	29,774	3.37
Mode 2	EF (energy) (gha/ton)	0.0307	47.7	0.033	47.76	0.031	47.76	0.031	47.7
	Total EF (gha/ton)	0.4873	5.45	0.506	5.669	0.499	5.402	0.495	6.04
	Cost (Rs/ton)	30,154	0.62	31,238	0.65	30,622	0.62	30,408	1.32
Mode 3	EF (energy) (gha/ton)	0.0087	85.1	0.009	85.12	0.008	85.12	0.008	85.1
	Total EF (gha/ton)	0.4653	9.71	0.483	10.09	0.476	9.623	0.473	10.2
	Cost (Rs/ton)	29,822	1.72	30,877	1.80	30,285	1.72	30,070	2.41

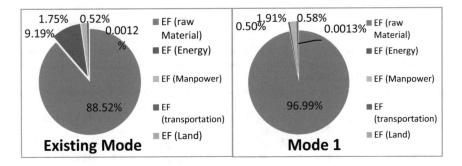

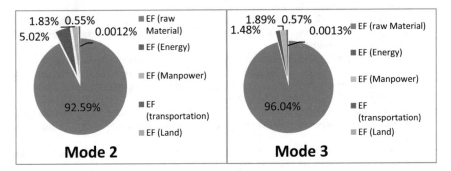

Fig. 17 Percentage EF share of resources and services for all the modes

the environmental impact as well as improve the economic savings due to cheaper sources of energy. Though the major impact of the industry comes from the raw material, energy utilization from renewable sources proved to improve the overall EF savings up to 8.8%.

The savings of EF of energy for three modes is estimated at about 95%, 47% and 85%, respectively. These savings of energy impacts contribute to the total EF savings of approximately 8.8%, 4.6% and 7.9%, respectively.

The cost savings by these three energy modes are about Rs. 407,963, Rs. 52,153.2 and Rs. 215,338.2, respectively. The environmental and economic savings together contribute to the improvement of ISI of the industry.

The ISI of the industry with the existing energy sources is calculated to be 0.81 units. By incorporating the suggested modes, ISI can be improved to 51.2, 1.53 and 9.23 for Mode 1, Mode 2 and Mode 3, respectively.

The usage of boiler-based steam for corrugation heating can be analysed based on the size of the corrugation plant. The steam-based usage reduces the maintenance of the gas burner and other allied problems.

The major impact of the industry comes from raw material, but the suggested modes are meant to reduce the environmental impacts of energy only. By reducing the raw material impacts drastic ecological as well as economic benefits can be attained.

References

1. BP Statistical World Energy Report (2018, June) Retrieved from https://www.bp.com/content/dam/bp/business-sites/en/global/corporate/pdfs/energy-economics/statistical-review/bp-stats-review-2018-full-report.pdf
2. Baidya S, Borken-Kleefeld J (2009) Atmospheric emissions from road transportation in India. Energy Policy 37:3812–3822
3. Barrera-Roldh A, Saldivar A, Ortiz S, Rosales P, Nava M, Aguilar S et al (2003) Industrial sustainability index. Trans Ecol Environ 63
4. Biswas A, Husain D, Prakash R (2020) Life-cycle ecological footprint assessment of grid-connected rooftop solar PV system. https://doi.org/10.1080/19397038.2020.1783719
5. Dietery Guidelines for Indians—a manual (2011) Retrieved from http://ninindia.org/: http://ninindia.org/DietaryGuidelinesforNINwebsite.pdf
6. Energy Statistics 2019 (n.d.) Retrieved 2020, from Ministry of Statistics and Program Implementation. http://www.mospi.gov.in/sites/default/files/publication_reports/Energy%20Statistics%202019-finall.pdf
7. Gielen D, Taylor P (2009) Indicators for industrial energy efficiency in India. Energy 34(8)
8. Global Footprint Network (GFN) (2010) Calculation methodology for the national footprint accounts. http://www.footprintnetwork.org/content/images/uploads/National_Footprint_Accounts_Method_Paper_2010.pdf. Accessed 21 Aug 2018
9. Global Footprint Network (GFN) (2017) Ecological wealth of nations. http://www.footprintnetwork.org/content/documents/ecological_footprint_nations/ecological.html. Accessed Dec 2018
10. Husain D, Garg P, Prakash R (2019) Ecological Footprint assessment and its reduction for industrial food products. Int J Sustain Eng
11. Husain D, Prakash R (2019) Ecological footprint reduction of built envelope in India. J Build Eng 21:276–286. https://doi.org/10.1016/j.jobe.2018.10.018
12. Indian Paper Industry (2018) Moving out of the woods? CARE Ratings Ltd
13. International Energy Outlook (2016) Retrieved from https://web.archive.org/web/20170727110053/https://www.eia.gov/outlooks/ieo/pdf/0484(2016).pdf
14. Pandey AK, Prakash R (2018) Industrial sustainability index and its possible improvement for paper industry. Open J Energy Effi
15. Pandey A, Prakash R (2020) Opportunities for sustainable improvement in aluminum industry. Engineering Reports
16. Pflug J, Verpoest I, Vandepitte D (1999) Folded honeycomb cardboard and core material for structural applications. EMAS
17. Quick Packaging News (n.d.) Retrieved from http://quickpakinc.blogspot.com/2016/06/corrugated-box-construction.html
18. Rocamora AM, Solis-Guzman J, Marrero M (2017) Ecological footprint of the use and maintenance phase of buildings: maintenance tasks and final results. Energy Build 155:339–351
19. Schmidt W-P, Taylor A (2006) Ford of Europe's product sustainability index. Retrieved from http://citeseerx.ist.psu.edu/viewdoc/download?doi=10.1.1.421.4459&rep=rep1&type=pdf
20. Shuaib M, Seevers D, Zhang X, Badurdeen F, Rouch KE, Jawahir IS (2014) A matrics-based framework to evaluate the total lifetime sustainability of manufacturing products. J Ind Ecol
21. Urbanik TJ (2001) Effect of corrugated flute shape on fibreboard edgewise crushing strength and bending stiffness. J Pulp Pap Sci 27:330–335
22. Wackernagel M, Rees W (1996) OUR ECOLOGICAL FOOTPRINT—reducing human impact on earth. New Society Publishers
23. World and China Energy Outlook 2050 (2018, November) Retrieved from https://eneken.ieej.or.jp/data/8192.pdf

Ascertainment of Ecological Footprint and Environmental Kuznets in China

Edmund Ntom Udemba

Abstract China's ecological footprint is considered to be the highest globally. This couples with the rising concern of climate change, the present study seeks to investigate the China's ecological impact with environmental Kuznets curve (EKC). China's country specific data of 1979–2018 was employed for this study. Scientific approaches such as autoregressive distributed lag (ARDL)-bound testing, VECM granger causality and some diagnostic tests were all employed for effective justification of the findings of this study. Positive relationships at both 1 and 5% significant level were established between ecological footprint and the explanatory variables. Environmental Kuznets curve was confirmed for China. Findings from granger causality confirmed nexus among the selected variables through both unidirectional and bidirectional transmission.

Keywords Ecological footprint · EKC · (ARDL)-bound · VECM granger causality · China

1 Introduction

China's ecological footprint is toping in the chart of global footprint (Global Footprint Network). The country's ecological footprint has dramatically increased for the past decade and as well grown more than that of United State of America (USA). China is among the most populated countries of the world. It is on record that China's ecological footprint ranked 81st in the globe at 2.5 gha which is considered less than the global average per capita ecological footprint at 2.7 but greater than the world average bio-capacity available per person at 1.7 gha. The tragedy of ecological footprint expansion is the resultant of overshoot which amounts to depletion of the available resources sustaining human lives (Global Footprint Network). This promotes activities such as diminishing of forest cover, depletion of fresh water

E. N. Udemba (✉)
Faculty of Economics Administrative and Social Sciences, Istanbul Gelisim University, Istanbul, Turkey
e-mail: eudemba@gelisim.edu.tr; eddy.ntom@gmail.com

© The Author(s), under exclusive license to Springer Nature Singapore Pte Ltd. 2021
S. S. Muthu (ed.), *Assessment of Ecological Footprints*,
Environmental Footprints and Eco-design of Products and Processes,
https://doi.org/10.1007/978-981-16-0096-8_3

systems, collapsing of fisheries, and build up of carbon dioxide emission. This by extension creates global climate change problem.

Testing ecological footprint and environmental Kuznets curve of China is essential considering the theoretical assertion of mitigating environmental damage with economic growth. Environment depletion is hypothesized to be corrected in the process of economic growth, hence the three stages of economic growth depict the trend of starting with overzealous act of promoting economic growth with less concern to the environment performance until it gets to a certain point of economic growth where the economy and environment will be given equal attention paving way for positive performance of economy and environment. China's economic growth is enormous with the massive ecological footprint which has remained a valid ground for analyzing ecological footprint and Kuznets curve.

Among the factors that impact ecological footprint are population, economic activities such as production that amounts to high utilization of energy sources [36], and agricultural activities like farming. Considering the definition of ecological footprint as a measuring tool of a population's use of resources which is linked to bio-capacity and a measureable volume of productive area [8], population is an inseparable influencer of ecological footprint. The continuous increase of population means the increase ecological footprint. This amount to recovering of forest covered areas by deforestation because of the need for farming and constructing shelter for the increasing population. The act of over utilizing resource from the ecology without consideration of the environmental impact poses danger of climate change [27]. Another indispensable factor that impact both ecological quality and economic growth is energy utilization. The impact of energy use is expected to be strong and positive moving toward greater economic growth through manufacturing and productive activities. The utilization of the energy (non-renewable) sources in manufacturing ventures impact the environmental quality negative through high emission from industrial activities and transportation sector [5]. Economic growth is another factor that impacts the ecological performance either positively or negatively. This depends on the model and policies adopted by the authorities of the economy.

This study is a country-specific investigation on China's ecological performance. Looking at the economic performance and the population of China, it is essential to consider the performance of the ecology and economic growth through the environmental Kuznets curve. It is no more news that China pose as the biggest consumer of energy as well as the biggest carbon emitter in the globe [5]. China's energy consumption increases from 0.57 billion tons of coal equivalent in 1978 to 4.49 billion tons of coal equivalent in 2017 (China Energy Statistics Yearbook). The increase in China's energy consumption has impacted the economic performance and as well caused some level of environmental problems. Different countries of the world have diverse ecological footprint with advanced countries taking measures to control their ecological footprint [33], while the developing countries are not matching with the same measure to save the ecology [3]. The manufacturing system of the developing nations is mostly on non-renewable energy sources that are not compatible with the quality of the environment [13].

On this note, considering the increased level of energy use and economic advancement of China the author wishes to investigate ecological footprint performance of China amidst massive energy utilization and population growth. In order to effectively execute this current study some important variables such as energy use, population, and agriculture are adopted to test the ecological footprint of China. Also, ecological footprint was adopted as proxy to the environment (ecology). Different scholars have investigated the position of China in global climate change using different variables such as carbon emission and greenhouse gas (ghg) emission. Ecological footprint has been proved to be the better option in measuring the environment performance because of its components such as carbon footprint, built-up land, cropland, grazing land, forest land, and fishing grounds [9, 11, 26, 38].

The rest sections of this study will be theoretical background, data and methodology, empirical analyses and result presentations, and conclusion and policy framing.

2 Literature and Theoretical Background

2.1 Literature Review

Several scholars have actually investigated ecological footprint and Kuznets with a combination of the selected variables. Al-Mulali et al. [2] studied ecological footprint with ecological Kuznets curve (EKC) and found EKC for the countries of researched on. Bagliani et al. [4] researched on the ecological footprint of Italy and found a breakeven total balance. Solarin and Bello [32] also researched the ecological footprint of 128 developed and developing economies and found non U-shaped of the EKC. Kivyiro and Arminen [14] investigated the carbon emission with economic growth and foreign direct investment, and they found EKC for the countries they researched on. Ozturk et al. [20] also researched EKC for the middle and higher income, and found negative relationship between ecological footprint and the explanatory variables. Yu-ming [42] studied the guangxi ecological footprint and energy use, and found a U-shaped curve ecological footprint. Neequaye and Oladi [19] researched on economic growth and foreign direct investment (FDI) and found EKC. Xu and Lin [41] researched on China carbon emission and found EKC for China. Ahmed and Long [1] studied Pakistan environmental performance and found EKC for Pakistan. Sarkodie and Strezov [28] researched on Brics and found EKC for India and South Africa. Shahbaz et al. [31] studied the implication of energy consumption on Bangladesh and found the selected variables impacting the positively on emission. Ullah et al. [37] investigate the implication of agricultural policy on environment and found agriculture impacting Pakistan environment.

2.2 Theoretical Background

Theoretically, this study is anchored on environmental Kuznets curve (EKC). EKC is a hypothesis coined from Kuznets curve of testing the income inequality among the farmers and white-collar workers. Simon Kuznets, [15]. He (Kuznets) tested the income inequality and came to conclusion that as rural farmers move to urban cities for white-collar jobs, the per capita income of the farmers who later joined the white-collar jobs will grow to meet the income of the rich at a point where curve exist. At this point, the income inequality is reduced. After establishment of this hypothesis by Kuznets, some energy and environmental economics [10, 21, 29] adopted this hypothesis to study the impact of economic growth on environment. It is hypothesized that as the economy grow the quality of the environment will be impacted negatively until it gets to a certain point where the masses will embrace the awareness of environmental performance and begin to work toward betterment of the environment. That particular point is called environmental Kuznets curve.

3 Data, Methodology, and Model Specification

A country specific data (1979–2018) of China is applied in this study. The selected variables applied for efficient investigation of the ecological performance in China are per capita of ecological footprint (*this comprises six components namely; carbon footprint, built-up land, cropland, grazing land, forest land, and fishing grounds*), GDP per capita (constant, 2010 U$), agriculture, forestry, and fishing, value added (constant 2010 US$), urban population and energy use (million tonnes oil equivalent). The energy use is the summation of different fossil fuel energy sources (oil; natural gas and coal, all measured in million tonnes oil equivalent) applied in China. Apart from ecological footprint per capita and energy use which are sourced from the Global Footprint Network and [25] British Petroleum (BP) statistical review, respectively, all other variables are sourced from World Bank Development Indicators (WDI). All series are converted to natural logarithm (Fig. 1).

Brief summary of data and variable is displayed in Table 1.

Methodologies adopted in this study are descriptive statistics, the test of unit root, autoregressive distributive lag (ARDL), diagnostic tests, and granger causality test. Descriptive statistics and unit root tests [6, 22]; Kwiatkowski-Philips-Schmidt-Shin [16] were employed to account for the stationarity and normality of the data, respectively, while ARDL-bound tests [23, 24] and granger causality were employed for cointegration, short and long run and for forecasting estimations. Also, diagnostic tests were employed for robust checking and the accuracy of the methodologies and techniques applied in this study.

The model specification is anchored on ARDL as proposed by [23, 24]. First, the econometric and empirical specifications are expressed as follow:

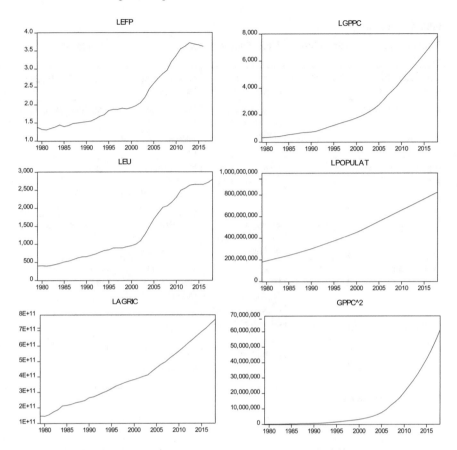

Fig. 1 Trend of variables in the data. Source Compiled by the author with Eviews

$$\text{EFP} = a_1 + a_2\text{GDP}_{it} + a_3\text{GDP}^2_{it} + a_4\text{EU}_{it} + a_5\text{AG}_{it} + a_6\text{POP}_{it} + \mu_{it} \quad (1)$$

$$\text{LNEFP} = a_1 + a_2\text{LNGDP}_{it} + a_3\text{LNGDP}^2_{it} + a_4\text{LNEU}_{it} + a_5\text{LNAG}_{it}$$
$$+ a_6\text{LNPOP}_{it} + \mu_{it} \quad (2)$$

where EFP, GDP, EU, AG, and POP from both Eqs. 1 and 2 mean ecological footprint, GDP per capita, energy use, agriculture and population, and a_1, a_2, a_3, a_4, a_5 and a_6 are the coefficients of the selected variables in the model. μ is the error term.

The model specification for short- and long-run analyses according to ARDL [23, 24] is expressed as following;

$$\ln \text{EFP}_t = \emptyset + a_1 \ln \text{EFP}_{t-1} + a_2 \ln \text{GDP}_{t-1} + a_3 \ln \text{GDP}^2_{t-1}$$

Table 1 Summary of variables

Definition of variables	Short form	Measurement	Source	Literature
Ecological footprint	LnEFP	Constant per capita	Global Footprint Network	[27, 2, 20, 34, 38]
Economic growth	LnGDP	GDP per capita (constant 2010, US$)	World Bank Development Indicators (WDI) [39]	Shahbaz et al., [31]
Squared economic growth	LnGDP^2	GDP per capita (constant 2010, US$)	World Bank Development Indicators (WDI), [39]	[14]
Energy use	LnEU	Energy use (million tonnes oil equivalent)	[25] British Petroleum (BP) statistical review	Shahbaz et al., [30]
Agriculture	LnAg	Agriculture, forestry, and fishing, value added (constant 2010 US$)	World Bank Development Indicators (WDI), [39]	[18] and Ullah et al., [37]
Population	LnPop	Urban population	World Bank Development Indicators (WDI), [39]	Wang et al., [40].

Source Author's compilation

$$+ a_4 \ln \mathrm{EU}_{t-1} + a_5 \ln \mathrm{AG}_{t-1} + a_6 \ln \mathrm{POP}_{t-1} + \sum_{i=0}^{p-1} \delta_1 \Delta \ln \mathrm{EFP}_{t-i}$$

$$+ \sum_{i=0}^{q-1} \delta_2 \Delta \ln \mathrm{GDP}_{t-i} + \sum_{i=0}^{q-1} \delta_3 \Delta \ln \mathrm{GDP}^2_{t-i} + \sum_{i=0}^{q-1} \delta_4 \Delta \ln \mathrm{EU}_{t-i}$$

$$+ \sum_{i=0}^{q-1} \delta_5 \Delta \ln \mathrm{AG}_{t-i} + \sum_{i=0}^{q-1} \delta_6 \Delta \ln \mathrm{POP}_{t-i} + \mathrm{ECM}_{t-i} + \mu_t \qquad (3)$$

Again, the variables from Eq. 3 have been explained except for some nomenclatures such as Δ, a_1, δ_1, ECM_{t-i}, and μ_t which denote first difference, short- and long-run coefficients, error correction, and white noise error term, respectively. Bound testing was applied for the cointegration analysis having confirmed the mixed order of integration in the unit root test estimate. In order to make a valid conclusion about the long-run relationship among the selected variables, hypothetical statements are made and expressed with the coefficients of long run. Hence, no cointegration is expressed as $H_0 = $ long-run coefficients (e.g., $a_1 = a_6 = 0$, $F -$ stat bounds), and there is cointegration or long-run relationship between the variables with alternative hypothesis as H_1 = long-run coefficients (e.g., $a_1 = a_6 0$, F − stat bounds). The F-statistics is compared with the critical values of lower and upper bounds. If the F-stats is greater, lesser, or in between the values of lower (I (0)) and upper (I (1)) bounds, the result is read as cointegrated, non-cointegrated, or inconclusive, respectively.

4 Empirical Results and Discussion

4.1 Descriptive Statistics

The output of descriptive statistics shows the variables with the highest variability as economic growth as represented with gross domestic product (GDP) per capita followed by energy use. The normality of the data is viewed with the output of kurtosis, skewness, and Jarque-Bera. From the output and the probability of the Jarque-Bera, it is evident that the data is normally distributed except for the GDP per capita that is significant at 5% which confirmed the high variability of the variable as stated earlier. With the level of normality existed in the data and model, linear analysis is considered more appropriate for the analysis (Table 2).

Table 2 Descriptive statistics

	LEFP	LGPPC	LEU	LAGRIC	LPOPULAT
Mean	2.187181	2233.916	1239.886	3.78E+11	4.39E+08
Median	1.878438	1489.627	894.8593	3.57E+11	4.13E+08
Maximum	3.720312	6907.962	2656.228	7.16E+11	7.82E+08
Minimum	1.306751	326.0476	396.3030	1.45E+11	1.80E+08
Std.Dev.	0.833841	1972.862	796.8581	1.66E+11	1.85E+08
Skewness	0.747463	1.004496	0.708679	0.461567	0.315711
Kurtosis	2.051145	2.746497	1.958637	2.141122	1.833070
Jarque-Bera	4.963956	6.492156	4.897786	2.517261	2.787331
Probability	0.083578	0.038927	0.086389	0.284043	0.248164
Sum	83.11286	84,888.81	47,115.65	1.44E+13	1.67E+10
SumSq.Dev.	25.72576	1.44E+08	23,494,363	1.01E+24	1.27E+18
Observations	38	38	38	38	38

Source Computed by the author

4.2 Unit Root/Stationarity Test

In every time series study, it is very essential to account for stationarity of the data. It is expected that the data are impacted by the structural events in the economy within the chosen period of research, and for this reason, stochastic trends are unavoidable in the history of the data which are likely to affect the outcome of the study if not accounted for. For this, unit root was tested in this study to ascertain the stationarity of the data and the order of integration. Author employed [6, 22]; Kwiatkowski-Philips-Schmidt-Shin [16] for effective estimation of the unit root, and unit root with mixed order of integration were found. Also, Chow test was applied for a robust check of the stationarity test. Most times, conventional tests are found to be weak in face of structural break caused by some natural or economic shocks (e.g., recession, pandemic, or epidemic) which always create a permanent shock to an economy capable of affecting the stationarity of data for the period of occurrence. With the aid of Chow tests, a structural break was identified in the year 2008, and dummy variable was created to correct the impact on the data. The outcomes of both the conventional tests Augmented Dickey Fuller [6]; Philip [22]; Kwiatkowski-Philips-Schmidt-Shin, Kwiatkowski et al. [16] (ADF, PP, and KPS) and the Chow tests are presented on Tables 3 and 4.

4.3 Test for Cointegration (Both the Short and Long Run) and Diagnostic Tests

ARDL-bound testing and other diagnostic tests were applied in this research for purpose of identifying the existence of cointegration (i.e., long-run relationship between the variables) and for confirming the stability, reliability, and fitting of the model. Bound testing has advantages over other methods of testing cointegration such as Engle and Granger [7] and Johansen [12]. Bound testing can be applied even when the series are integrated of mixed order or on the same order, and also, it can be applied when the sample is small. The result of the cointegration (*short-run and long-run relationships that exist between the selected variables*) estimation and the diagnostic tests are presented in Table 5. The outputs on the table are explained as follows: The goodness of fit which showed the part of dependent variable (lnEFP) that are explained by the independent variables (lnEU, lnGDP, $lnGDP^2$, lnAG, and lnPOP) is represented with $R^2 = 0.999635$ and adjusted $R^2 = 0.999327$ while the remaining part of dependent variable is explained by residual (error term $= \mu$). Auto-correlation and serial correlation are tested with Durbin Watson and Breuch–Godfrey serial correlation LM tests. From the outputs of both tests absence of autocorrelation or serial correlation is confirmed with the values of Durbin Watson (2.22) and serial correlation LM tests (F-stats $= 0.5734$ and Chi-square $= 0.4117$). Among the results displayed on the table is the cointegration from bound testing. With the values of F-stats (6.320642) and critical values of both lower (4.045) and upper (5.898) bounds,

Table 3 Stationarity test: ADF, PP, and KPS tests of unit root

Variables		@ LEVEL		1st Diff		
	With intercept	Intercept and trend	With intercept	Intercept and trend	Decision
PP						
LEFP	0.8996	−1.8391	−2.7176**	−2.4810	I(1)
LGDP	9.1700	1.9461	0.2469	−2.2934	I(1)
LEU	0.9246	−1.7294	−2.3476	−2.5640	I(1)
LAG	6.0466	1.0448	−3.4803***	−5.0491***	I(1)
LPOP	7.0495	−2.6048	−1.4903	−0.6896	I(2)
ADF						
LEFP	−0.4482	−2.1887	−2.7035**	0.3590	I(1)
LGDP	1.2546	0.4060	0.0529	0.4533	I(1)
LEU	0.0720	−2.1067	−2.2944	0.3734	I(1)
LAG	6.8049	1.2967	−1.8071	0.0017**	I(1)
LPOP	−0.7129	−2.7767	−1.5396	−2.2431	I(2)
KPSS						
LEFP	0.6757**	0.1804**	0.3264	0.1015	
LGDP	0.7051**	0.2046**	0.7129**	0.1642**	
LEU	0.7178**	0.1785**	0.3272	0.0992	
LAG	0.7663***	0.1969**	0.7378**	0.1221*	
LPOP	0.7713***	0.2069**	0.7108**	0.1634**	

Notes (a): (*) Significant at the 10%; (**) Significant at the 5%; (***) Significant at the 1%(b): *P*-value according to (1) Maclean et al. [17] one-sided *P*-values (2) Kwiatkowski-Phillips-Schmidt-Shin [16]

Table 4 Chow breakpoint test: 2008

F-statistic	9.387994	Prob. F(6,26)	0.0000
Log likelihood ratio	43.79934	Prob. Chi-Square(6)	0.0000
Wald statistic	56.32796	Prob. Chi-Square(6)	0.0000

cointegration is confirmed even at 1%. The optimal lag is estimated and found to be 1 with Akaike Information Criterion (AIC). Error correction model (ECM) confirmed the speed of adjustment at 0.419 (−0.4193) and significant at 1%. This shows the ability of the model output to restore equilibrium when structural events or disequilibrium is noticed, and the long-run relationship among the variables are established. The reliability and stability of the model are as well tested with cumulative sum and cumulative sum square (CUSUM &CUSUM^2). At first, it was found that there is instability which was corrected by creating and estimated with dummy variable. The

Table 5 Cointegration (ARDL) assessments of EFP model

Variables	Coefficients	SE	t-statistics	P-value
Long-path				
LGDP	0.000348	0.000138	2.5235	0.0187**
LGDP^2	−1.09E−07	3.95E−08	−2.7468	0.0112**
LEU	0.001188	0.000180	6.5889	0.0007***
LAG	4.98E−12	1.36E−12	3.6582	0.0012***
LPOP	2.55E−09	1.07E−09	2.3689	0.0262**
Constant	0.51894	0.18089	2.8687	0.0085***
Short-path				
D(LGDP)	0.000348	0.000138	2.5235	0.0187**
D(LGDP^2)	−1.09E−07	3.95E−08	−2.7468	0.0112**
D(LEU)	0.00119	0.000108	11.0272	0.0000**
D(LAG)	4.98E−12	9.97E−13	4.9892	0.0000***
D(LPOP)	2.55E−09	1.07E−09	2.3689	0.0262**
CointEq(−1)*	−0.41931	0.06194	−6.7694	0.0000***
R^2	0.999248			
Adj.R^2	0.998903			
D.Watson	2.2215			
Bound test (Long-path)				
F-statistics	6.3206***	$K = 5, @ 1\%$	I(0)bound = 4.257	I(1)bound = 5.898
Wald test (short-path)				
F-statistics	2898.565***			
P-value	0.00000***			
Serial correlation test				
F-statistics	0.5704			
Chi-square	0.4117			
P-value	0.5734			
Heteroscedasticity test				
F-statistics	0.49957			
Chi-square	0.8223			
P-value	0.8848			

Note *, **, *** Denotes rejection of the null hypothesis at the 1, 5 and 10%
Sources Authors computation

outputs are presented immediately after Table 5. The short- and long-run relationships between the dependent (ecological footprint) and the independent (energy use, agriculture and population) variables as displayed on the table are interpreted and explained as follows: environmental Kuznets curve is confirmed for China with positive and negative relationships that existed among GDP per capita, squared GDP per capita, and ecological footprint. At first, a positive relationship is established between GDP per capita and ecological footprint at 5% significant level depicting increase in economic growth is overshooting ecological footprint, but a negative relationship existed between squared GDP per capita and ecological footprint at 5% significant level. This scenario supports the Kuznets hypothesis of straining the ecology at the expense of economic growth. Numerically, a percentage increase in GDP per capita will lead to 0.000348 increase of ecological footprint, while for squared GDP per capita, a percent increase in economic growth will lead to 0.00000109 degrease in ecological footprint. This supports the findings by Kivyiro and Arminen [14]. Positive relationship is established between energy use and ecological footprint both in short and long run. This implies a percent increase in energy use will lead to 0.001188% increase in ecology (poor environment). This finding is in consonance with the finding of Udemba et al. [35] for Indonesia. Also, for the case of agriculture and population, positive relationship is established both in short and long run for the both variables. A percent increase in agriculture and population will, respectively, lead to 0.0000000000498 (4.98E−12) and 0.0000000255 (2.55E−09) increase in ecological footprint (poor environment). These findings support the findings by udemba [36] for India; Liu et al. [18] and Ullah et al. [37]; Wang et al. [40]. In summary, the output of the cointegration with regard to short run and long run shows that the selected variables are really impacting ecology unfavorably.

4.4 Diagnostic Tests (CUSUM and CUSUM²)

See Figs. 2, 3, 4 and 5.

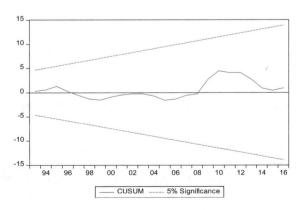

Fig. 2 CUSUM residual graphical plot before structural break test with Chow

Fig. 3 CUSUM^2 residual graphical plot before structural break test with Chow

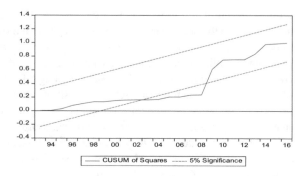

Fig. 4 CUSUM residual graphical plot after the structural break test with Chow

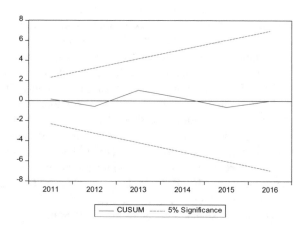

Fig. 5 CUSUM^2 residual graphical plot after the structural break test with Chow

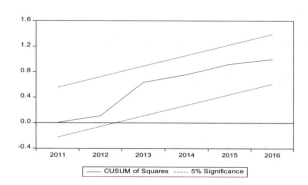

4.5 Granger Causality Analysis (VECM)

Granger causality is applied in this study for a robust check to the cointegration estimates and for confirmation of the forecasting power of the variables. The cointegration analysis confirms the level of relationship that exists between the dependent and independent variables without much insight as to which variable is inducing or

Table 6 Short- and long-run VECM granger causality analysis/block exogeneity wald tests

Variables			Long run		
	LEFP	LGDP	LEU	LAG	LPOP
LEFP	✓✓ ✓✓	0.003 [0.96]	0.866 [0.35]	0.242 [0.62]	4.57 [**0.03**]
LGDP	2.586 [0.10]	✓✓ ✓✓	2.132 [0.14]	0.177 [0.67]	5.615 [**0.02**]
LEU	2.619 [0.10]	5.667 [**0.02**]	✓✓ ✓✓	1.374 [0.24]	0.364 [0.55]
LAG	11.10 [**0.00**]	2.827 [**0.09**]	12.41 [**0.00**]	✓✓ ✓✓	19.70 [**0.00**]
LPOP	17.60 [**0.00**]	4.926 [**0.03**]	20.17 [**0.00**]	2.409 [0.12]	✓✓ ✓✓

Note **Bolden figures** in brackets are the prob. that represent 10, 5, and 1% significance resp. while the figures before the brackets are the Chi-squares (χ^2) = Chi-squares (χ^2) [P-values]

impacting the other. The granger causality analysis gives clear direction of the relationship that exists among the variables, whether it is unidirectional or bidirectional casual relationship. In this study, block exogeneity vector error correction model (VECM) granger causality is utilized in estimating the granger causality relationship among the variables, and the result is shown in Table 6.

From the output of the granger causality, a unidirectional transmission exists between ecological foot print and agriculture (i.e., agriculture is transmitting to ecological footprint), unidirectional transmission is seen passing from both energy use and agriculture to economic growth, likewise from agriculture and population to energy use, and from agriculture to population. Bidirectional is seen between ecological footprint and population, between economic growth and population. The output supports both the population and agricultural-induced ecological footprint as equally recorded in the cointegration output. Also, with energy use impacted by agriculture and population, economic growth impacted by energy use and agriculture, there is a nexus among the selected variables. This means that the variables can forecast themselves.

5 Conclusion and Policy Recommendation

The present study focuses on the assessment of ecological footprint and environmental Kuznets curve for China. Much has been said about China with respect to carbon emission and greenhouse gas emission with little emphasis on ecological footprint. Also, considering the economic performance of China with respect to industrial activities and economic growth which utilize excessive fossil fuel energy and the position of China in a global population, it is essential to research on the likelihood of the country overshooting the ecological footprint.

Scientific approaches such as autoregressive distributed lag (ARDL)-bound testing, vector error correction model (VECM) granger causality and some diagnostic tests were all employed for effective justification of the findings of this study, detailed in the data and methodology sections. Positive relationships at both 1 and

5% significant level were established between ecological footprint and the explanatory variables. Environmental Kuznets curve was confirmed for China. Findings from granger causality confirmed nexus among the selected variables through both unidirectional and bidirectional transmission.

The policy framing should focus on encouraging economic growth through clean production as EKC is confirmed for China. China should consider adopting more renewable energy such as wind, solar, and other renewable energies. The beacon of environmental Kuznets curve emphasized the positive implication of continuous economic growth on environment. Considering the continuous growth of China's economy, the environment performance of China will be impacted positively. Again, transition to a more conservative energy sources such as wind, solar, and hydropower will effectively correct the environment degradation of China. It is evident that population is among the determinant of the ecological footprint and the economic growth of China, therefore, policy framing should consider standardizing of the population of the country to avoid overshooting of ecological footprint.

Conclusively, this study and the policies recommended are significant to the emerging countries that have similar features with China.

Funding Author wishes to inform the Editor/Journal that no form of funding was received for this research.

Compliance with Ethical Standards
Author wishes to inform the Editor/Journal that there are no conflicts of interest at any level of this study.

References

1. Ahmed K, Long W (2012) Environmental Kuznets curve and Pakistan: an empirical analysis. Proc Econ Finance 1:4–13
2. Al-Mulali U, Weng-Wai C, Sheau-Ting L, Mohammed AH (2015) Investigating the environmental Kuznets curve (EKC) hypothesis by utilizing the ecological footprint as an indicator of environmental degradation. Ecol Ind 48:315–323
3. Aremo AG, Ojeyinka TA (2018) Foreign direct investment, energy consumption, carbon emissions and economic growth in Nigeria (1970-2014): an aggregate empirical analysis. Int J Green Econ 12(3–4):209–227
4. Bagliani M, Galli A, Niccolucci V, Marchettini N (2008) Ecological footprint analysis applied to a sub-national area: the case of the Province of Siena (Italy). J Environ Manage 86(2):354–364
5. Denolle MA, Fan W, Shearer PM (2015) Dynamics of the 2015 M7. 8 Nepal earthquake. Geophys Res Lett 42(18):7467–7475
6. Dickey DA, Fuller WA (1979) Distribution of the estimators for autoregressive time series with a unit root. J Am Stat Assoc 74(366a):427–431
7. Engle RF, Granger CW (1987) Co-integration and error correction: representation, estimation, and testing. Econometrica: J Econometric Soc 251–276
8. Gatti RC, Callaghan TV, Rozhkova-Timina I, Dudko A, Lim A, Vorobyev SN, Pokrovsky OS (2018) The role of Eurasian beaver (Castor fiber) in the storage, emission and deposition of carbon in lakes and rivers of the River Ob flood plain, western Siberia. Sci Total Environ 644:1371–1379

9. Grimes P, Kentor J (2003) Exporting the greenhouse: foreign capital penetration and CO? Emissions 1980 1996. J World-Syst Res 261–275

10. Grossman GM, Krueger AB (1991) Environmental impacts of a North American free trade agreement (No. w3914). National Bureau of economic research

11. Halicioglu F (2009) An econometric study of CO_2 emissions, energy consumption, income and foreign trade in Turkey. Energy Policy 37(3):1156–1164

12. Johansen S (1988) Statistical analysis of cointegration vectors. J Econ Dyn Control 12(2-3):231–254

13. Kissinger M, Rees WE (2010) An interregional ecological approach for modelling sustainability in a globalizing world—reviewing existing approaches and emerging directions. Ecol Model 221(21):2615–2623

14. Kivyiro P, Arminen H (2014) Carbon dioxide emissions, energy consumption, economic growth, and foreign direct investment: Causality analysis for Sub-Saharan Africa. Energy 74:595–606

15. Kuznets S, Murphy JT (1966) Modern economic growth: rate, structure, and spread (Vol. 2). New Haven: Yale University Press

16. Kwiatkowski D, Phillips PC, Schmidt P, Shin Y (1992) Testing the null hypothesis of stationarity against the alternative of a unit root: how sure are we that economic time series have a unit root?. J Econometrics 54(1-3):159–178

17. Landry D, MacLean G (1996) The Spivak Reader. Routledge

18. Liu X, Zhang S, Bae J (2017) The impact of renewable energy and agriculture on carbon dioxide emissions: Investigating the environmental Kuznets curve in four selected ASEAN countries. J Clean Prod 164:1239–1247

19. Neequaye NA, Oladi R (2015) Environment, growth, and FDI revisited. Int Rev Econ Finance 39:47–56

20. Ozturk I, Al-Mulali U, Saboori B (2016) Investigating the environmental Kuznets curve hypothesis: the role of tourism and ecological footprint. Environ Sci Pollut Res 23(2):1916–1928

21. Panayotou T (1997) Demystifying the environmental Kuznets curve: turning a black box into a policy tool. Environ Dev Econ 465–484

22. Perron P (1990) Testing for a unit root in a time series with a changing mean. J Bus Econ Stat 8(2):153–162

23. Pesaran MH, Shin Y (1998) An autoregressive distributed-lag modelling approach to cointegration analysis. Econometr Soc Monogr 31:371–413

24. Pesaran MH, Shin Y, Smith RJ (2001) Bounds testing approaches to the analysis of level relationships. J Appl Econometr 16(3):289–326

25. Petroleum B (2019) BP statistical review of world energy report. BP, London, UK

26. Rees W, Wackernagel M, Testemale P (1996) Our ecological footprint: reducing human impact on the Earth. New Society Publishers, Gabriola Island, BC, pp 3–12

27. Rees WE (1996) Revisiting carrying capacity: area-based indicators of sustainability. Popul Environ 17(3):195–215

28. Sarkodie SA, Strezov V (2019) Effect of foreign direct investments, economic development and energy consumption on greenhouse gas emissions in developing countries. Sci Total Environ 646:862–871

29. Shafik N, Bandyopadhyay S (1992) Economic growth and environmental quality: time series and cross section evidence. Policy research working paper N° WPS904, World Bank

30. Shahbaz M, Nasreen S, Ahmed K, Hammoudeh S (2017) Trade openness–carbon emissions nexus: the importance of turning points of trade openness for country panels. Energy Econ 61:221–232

31. Shahbaz M, Uddin GS, Rehman IU, Imran K (2014) Industrialization, electricity consumption and CO_2 emissions in Bangladesh. Renew Sustain Energy Rev 31:575–586

32. Solarin SA, Bello MO (2018) Persistence of policy shocks to an environmental degradation index: the case of ecological footprint in 128 developed and developing countries. Ecol Ind 89:35–44

33. Solarin SA, Al-Mulali U (2018) Influence of foreign direct investment on indicators of environmental degradation. Environ Sci Pollution Res 25(25):24845–24859
34. Uddin GA, Salahuddin M, Alam K, Gow J (2017) Ecological footprint and real income: panel data evidence from the 27 highest emitting countries. Ecol Ind 77:166–175
35. Udemba EN, Güngör H, Bekun FV (2019) Environmental implication of offshore economic activities in Indonesia: a dual analyses of cointegration and causality. Environ Sci Pollut Res 26(31):32460–32475
36. Udemba EN (2020) Mediation of foreign direct investment and agriculture towards ecological footprint: a shift from single perspective to a more inclusive perspective for India. Environ Sci Pollut Res 1–18
37. Ullah A, Khan D, Khan I, Zheng S (2018) Does agricultural ecosystem cause environmental pollution in Pakistan? Promise and menace. Environ Sci Pollut Res 25(14):13938–13955
38. Ulucak R, Lin D (2017) Persistence of policy shocks to Ecological Footprint of the USA. Ecol Ind 80:337–343
39. WDI T (2019) World development indicators (DataBank)
40. Wang Y, Kang L, Wu X, Xiao Y (2013) Estimating the environmental Kuznets curve for ecological footprint at the global level: a spatial econometric approach. Ecol Ind 34
41. Xu B, Lin B (2015) Factors affecting carbon dioxide (CO_2) emissions in China's transport sector: a dynamic nonparametric additive regression model. J Clean Prod 101:311–322
42. Yu-ming WU (2010) Kuznets curve analysis of guangxi ecological footprint and energy consumption. China Popul Resour Environ 20(11):30–35

Development of Renewable Energies and Its Consequences on the Ecological Footprint

B. Senthil Rathi and P. Senthil Kumar

Abstract Fossil fuels focused on gas, oil, and coal are very useful for developing the national economy, however on the other side, a few of the negative environmental effects of these commodities have led us to use such commodities under some limits and to switch our attention toward renewable resources. In the near future, green energies will be an essential form of energy production, so we could utilize such sources repeatedly to generate usable energy. Statistical evidence helps in the long term, green energies and real production have a bad influence on the ecological footprint, whereas economic success and globalization have a beneficial impact on the ecological footprint. Short-term findings show that green energies, financial growth, real production, and globalization are positively related to the environmental footprint. This chapter summarizes the renewable energy types, importance of renewable energy, ecological footprint, and the relationship between the renewable energy and the ecological footprint. Furthermore, it has also concluded that rising renewable energy use decreases the environmental footprint and rising non-renewable resource use enhances ecological damage.

Keywords Fossil fuels · Ecological footprint · Environment · Renewable energy · Green energy

1 Introduction

Today, use of fossil energy is growing rapidly, together with changes in life quality, industrial development of developed nations, and a growth in the global population. These have traditionally been known that this unsustainable use of fossil fuels not only contributes to a rise in the rate of depletion in fossil fuel stocks, but also has a

B. Senthil Rathi
Department of Chemical Engineering, St. Joseph's College of Engineering, Chennai 600119, India

P. Senthil Kumar (✉)
Department of Chemical Engineering, Sri Sivasubramaniya Nadar College of Engineering, Chennai 603110, India
e-mail: senthilchem8582@gmail.com; senthilkumarp@ssn.edu.in

© The Author(s), under exclusive license to Springer Nature Singapore Pte Ltd. 2021
S. S. Muthu (ed.), *Assessment of Ecological Footprints*,
Environmental Footprints and Eco-design of Products and Processes,
https://doi.org/10.1007/978-981-16-0096-8_4

substantial negative effect on the atmosphere, resulting in heightened health threats and a challenge to climate change [10, 18, 26]. Increased use of fossil fuels to satisfy existing energy demands concern about the energy shortage has created a revival in interest in encouraging green alternatives to satisfy the growing energy challenges of the modern world [5, 69]. Too much use of non-renewable resources has induced global warming by CO_2; thus, the development of renewable energy from sustainable energies is desperately needed [2, 21, 36]. Changes in ecosystem development are growing more and more socially appropriate globally, particularly in developed nations. Society is increasingly shifting to more efficient development practices, minimization of waste, removal of automotive environmental pollution, distribution of power production, protection of forest areas, and reduce greenhouse gas emissions [37, 54].

Renewable energies are known as green energy technology options and the efficient usage of these products minimizes environmental effects, creates minimal ancillary emissions and are viable based on present and emerging socioeconomic empowerment. Renewable energy solutions offer an outstanding chance to minimize emissions of greenhouse gases and reduce global warming by eliminating traditional energy sources [44]. Clean energy supplies can play a key role in the sustainability of the planet. Energy resources have been categorized into three subgroups: non-renewable sources, natural energy, and radioactive energy [11]. Clean energy options comprise organic matter, hydroelectric power, solar, geothermal, tidal energy, wind energy, and ocean. Renewable electricity is the predominant, household and green or boundless energy option [13]. Renewable sources that satisfy household energy needs have the ability to produce nil or almost nil emissions of all atmospheric air pollutants and greenhouse emissions. Developing a clean energy infrastructure would make it easier to meet the most critical tasks at current such as enhancing the security of energy sources, biofuel currency, addressing regional electricity, water shortage problems, enhancing the effectiveness of life and jobs of the native community. It also maintains the sustainable economic growth of isolated areas in the desert and mountain areas and the introducing initiatives [14, 35, 70].

The excavation and utilization of natural resources intensify at the same time as a rise in profits, which leads to a reduction in biodiversity thus raising the ecological footprint [43, 63]. The ecological footprint has identified as a measure of ecological destruction in published studies [3, 7, 63]. Increasing economic development accelerates the exploitation of natural resources and raises the ecological footprint [43]. Even so, renewables are plentiful and viable, while non-renewable resources are wasteful and limited, and their use raises the ecological footprint [42]. Ecological footprint is commonly used in favor of sustainable assessment that could be built for evaluation and monitoring of commodity utilization throughout the world and evaluates the sustainable development of people's lives, products or services, businesses, urban centers, communities, regions, and nations [57]. Based on the above, it is indeed essential to remember the hypothesized relationship between total wages, leasing of natural resources and ecological footprint, taking into account possible variables such as sustainable energy and urbanization.

This chapter summarizes the forms of sustainable energy, the value of clean energy, the ecological footprint, and the interaction among renewable energy and the ecological footprint. It also reported that the usage of renewable energy lowers the environmental footprint and increasing use of non-renewable resources increases environmental harm.

2 Renewable Energy and Its Types

Renewable energy sources are energy supplies that are continuously refilled by natural world and generated explicitly or implicitly from the environment or even other natural motions and processes of the environment. Renewables do not really contain non-renewable energy supplies, fossil fuel by-products, or synthetic by-products [24, 33, 52]. The capacity for sustainable energy supplies is immense, and that they will, in theory, massively surpass the worldwide energy requirements; thus these resources would have a large part to play in the upcoming world energy supply, most of which is still centered on increasing their inventory of renewable sources [16]. Types of renewable energy are solar energy, wind energy, geothermal energy, biomass energy, and hydrogen as a fuel.

2.1 Solar Energy

Solar electricity is very inexpensive relative to many other sources of energy. They also were plentiful and ideal for a wide range of applications. Maintenance costs for solar power are also minimal. The key disadvantage is that they have been susceptible to intermittent climate; thus, the need for a power storage facility that contributes to the total expense of the technology [46, 68]. The development of sun energy grew rapidly around 30 years. These have developed from simple systems to mass energy. Since the invention of solar panels in the nineteenth century, many nations have used solar electricity production. The first was the USA, preceded by Japan and Germany, however currently China is the largest source of energy from the solar radiation [29]. Solar power has been used either in sunlight thermal systems whereby solar power is being used as a source of heat or partially used as a production of power in concentrated solar farms specifically used to create power in photovoltaic and concentrated solar photovoltaic systems [19, 47, 72].

Solar cooking has been the most quick and effective use of solar power. Solar power is a viable alternative worthy of becoming one of the pioneering energy sources for preparing food. Numerous kinds of solar cookers were affordable, from which solar cooker form box is commonly used in the globe [8, 41].

Solar water heater of daily level, sufficient to fulfill much of the warm water requirements of a family with children, provides substantial environmental security and should be used wherever necessary to ensure a better tomorrow [65]. It is projected that a household solar water-heating device with a volume of 100 L per day will minimize CO_2 emissions by approximately 1237 kg per year with an ability consumption rate of 50% and is approximately 1410.5 kg in sunny and warm areas [34].

Conversion process of sun energy to electrical energy is traditionally achieved utilizing photovoltaic systems that enable use of photovoltaic effect [48]. The photovoltaic effect relies on the association of light with radiation equal to or higher than the spectrum of photovoltaic materials. Any hindered due to band-gap limits are eliminated by rippling semiconductors with various band-gaps [20, 45]. Photovoltaic systems produce energy straight from sunlight no combustion, sound, or motion. Sun's rays is free, but the cost of producing electricity is incredibly high, while rates are beginning to decline. Solar power has a low density: Photovoltaic systems need a massive area for tiny amounts of energy production. The major ingredient of grid-connected photovoltaic panels is the boost converter, which transforms DC power into AC power along with the voltage and current quality specifications of the electricity network [55, 61].

2.2 Wind Energy

Wind power is characterized by transforming wind energy from wind turbines to usable forms, such as the use of wind generators for power generation, wind generators for mechanical power generation, wind pump for water pumped or irrigation, or sails for powering ships. The very first wind farms for the production of electricity were built at the start of the twentieth century. Since before the early 1970s, innovation has steadily changed. Wind energy has re-emerged as among the most popular renewable energy options since the late 1990s [32, 23, 51]. Producing wind power demands that perhaps the kinetic energy of flowing air be transformed to mechanical then somehow electric power, thereby requiring the technology to build budget wind farms and power stations to achieve this transformation. The sum of kinetic energy of motion, which is potentially usable for retrieval, rises with the wind velocity cube. However, the turbine absorbs just a portion of the usable energy (40–50%), so the architecture of the turbine focuses on optimizing energy recovery across the spectrum of wind gusts encountered by wind generators, while at the same time minimizing the cost of wind power for all variables. In order to decrease costs, the design of the wind turbine is often driven by a desire to reduce the use of components while increasing the size of the turbine, increasing part and device efficiency, and enhancing the operation of the wind turbine [17, 28].

In the power-hungry developing nations, wind energy is a feasible source of energy that can be built and delivered quite quickly, including in rural, remote, and steep hills regions. Power generation from the wind has not ever depletes and still never raises its cost. The energy generated by such schemes might save numerous billion barrels of oil and prevent numerous millions of tons of carbon and other pollutants [6, 56]. At an average wind speed of 4.5 m/s, the projected annual net CO_2 pollution reduction capacity is the minimum (2874 kg) for the GM-II system and the maximum (7401 kg) for the SICO framework in the current diesel replacement. Likewise, in the context of energy replacement at the very same sustained winds, it is calculated to be 2194 and 5713 kg, respectively, for the 2 scenario mentioned earlier [34].

2.3   Geothermal Energy

Geothermal energy is an effective and reliable technique for removing green energy again from planet by natural phenomena. It can be achieved on a microlevel to supply heat to a single home with a geothermal heating system or on a massive scale to generate electricity from a geothermal plant. Geothermal energy is known to be a cost-effective, safe, and environmentally sustainable source of energy [4, 60].

Geothermal energy can be considered sustainable at the period of production technologies that do not need the evolutionary time of fossil fuel deposits, like oil, coal, and gas. A treatment of large enthalpy storage is carried out around the same location from where the liquid or temperature is collected. Furthermore, renewable development can be done in duplication and heating systems. In particular, the environmental consequences of geothermal energy production and primary use can be minimal, manageable, or zero. The environmental standards, that may vary widely, should be strictly complied with. In either event, the results must be tracked and recorded (over much extended periods), graded although if possible, minimized [49].

Geothermal forms of energy are known as hydrothermal, conductive, and deeper groundwater networks. Geothermal processes have forms overtaken by liquids and vapors. Conductor's processes have hot stone and molten over even a broad range of conditions, and deeper aquifers include moving liquids in permeable materials or crack areas at depth usually greater than 3 km, but this shortage a concentrated magmas heating element. Geothermal power resource utilization systems can be categorized into forms for electricity production, hot air use or mixed power and heat in cogeneration implementations. Geothermal heat pump technology is a straightforward category. Actually, hydrothermal energy supply and specific usage are also the only economically utilized geothermal technologies [53]. Numerical models were planned to evaluate the recovery period for the key utilization schemes [50]. The estimated capacity installed of the geothermal energy is anticipated to be somewhere between 140 and 160 GWe by 2050, whereas the future planned potential for straightforward use may exceed 800 GWth [15].

2.4   Biomass Energy

Biomass seems to be the word associated with a particular biomaterial derived from plants, trees and grains, and seems to be the accumulation and storing of solar energy by photosynthetic activity. Bioenergy is the transformation of organic matter into usable sources of heat energy, electric power and liquid fuel sources. Organic matter for biomass energy arises mostly primarily from soil, like committed crop residues, or from residues produced in the cultivation of crops for food and other goods. Bioenergy is green and clean, but has several features in common with non-renewable. Although organic matter could be fired directly to produce electricity, it could also act as a raw material for conversion to different fluid fuels. Biofuels could be shipped and processed and provide on-demand power and heat production, which is important in an electricity sector that is heavily dependent on unreliable. These parallels reflect the main role that organic matter is destined to perform in sustainable energy situations [59].

Biogas processing by anaerobic digestion provides huge benefits over all other modes of biofuel production. This one has been assessed as among the most energy-efficient and ecologically responsible innovations for the development of biomass energy [66]. Many other various raw goods and digestive system processes can be used for the biogas production. Bioenergy has obvious benefits, particularly if opposed to many other clean energy options. This can be generated as required and therefore can be processed comfortably. This can be delivered via the current fossil fuel system and is used in the similar purposes as fossil fuels. Biogas could be specifically used only for home heating, shipping power or transmitted via the natural gas for final usage [25, 58].

Biodiesel is environmentally friendly, with lower CO_2 and NO_x pollutants. Consistent usage fossil fuel products is now generally accepted as unaffordable due to the scarcity of reserves and the exposure of such products to the concentration of atmospheric carbon dioxide. Clean energy, zero carbon, transportation energy sources are important for sustainable development. Algae also appeared as among the leading destinations for the manufacture of biofuel. This can be interpreted that organisms matured in CO_2 enriched air may be transformed to oils. A really strategy will help to address the big challenges of air emissions arising from the development of CO_2 gas and the potential crisis leading to a shortage of energy resources [27].

In comparison with the advantages, there have been major obstacles to biofuel production installations. Biofuels have lower power concentrations, and compilation and shipping can be costly. The use of biofuels to produce energy is technically well known, however the amount offered for electric power rarely negates the true amount of biomass resource. Biomass energy sources are strenuous in the need for components, including soil, water, plants, and coal power, all of whom have a relative value [44].

2.5 Hydrogen as a Fuel

Hydrogen has intrigued centuries of people over the centuries. Hydrogen is anticipated to lead role in the global health of the nation by eliminating non-renewable resources. Hydrogen is becoming increasingly necessary as an opportunity for potential electricity. The transformation into energy or electricity is quick and clean. If hydrogen is burning with oxygen, it releases no contaminants, just water that can contribute to natural environment. Even so, hydrogen, utmost abundant chemical compound on the earth, may not occur in its elementary earth's crust. It must be isolated from organic chemicals by electrolysis from water or by chemical reactions from petroleum products and perhaps other hydrogen reservoirs. Electric power for electrolysis may soon return from green energy such as solar radiation, air and water change in momentum, or geothermal. Hydrogen can therefore become a significant connection between green physical resources and energy source carriers [22, 30, 62].

The biggest problem is the successful conversion of renewable energies into power generation and the conservation of power or the manufacture of artificial fuel. Hydrogen is derived from water by means of an electrolysis cell. The preservation of hydrogen from its atomic or molecular type is a product problem. Many hydrides are considered to have a hydrogen capacity similar to oil; but these hydrides need a complex storage device. That power density of the device is slightly lower than that of the power output of non-renewable sources. Synthesized hydrocarbons formed from hydrogen and CO_2 removed from the atmosphere are an attractive possibility to direct hydrogen storage. These are CO_2 neutral and processed as non-renewable sources. Standard internal combustion and turbines could be used to transform the energy stored to work and energy. A big shift in the energy revolution from coal and natural gas carriers to renewable energy flows is required [73].

3 Importance of Renewable Energy

Renewable energy is a carbon-free or carbon-neutral power source, the expanded usage of which could be useful in several ways of programs and strategies implemented by certain industrialized and developing economies to fulfill their pollution obligations as laid out again in the Kyoto Protocol. Excess production of clean power will also assist governments accomplish certain key policies, like power protection and sustainability. Clean energy includes a broad range of types of fuel and consumer applications. A validity of each of these techniques and the expense of various renewable technologies differ tremendously. Any renewable energy, like organic matter, are commonly used and freely accessible. Some such as geothermal are commonly seen how accessible. Any alternative sources such as solar and wind have still not been shown on a wide scale, but they have an immense unexploited capacity. However, at the other side, the decreasing investments in renewable power, the enhancement of renewable energies as well as the involvement of government bodies in improving energy efficiency due to various their beneficial environmental consequences should result in continuous best growth with the use of sustainable power. Use of sustainable power has indeed worked to minimize the development in greenhouse gases from the energy industry, and governments around the world perceive enhanced use of sustainable power both as viable and beneficial [9].

There seem to be a range of research challenges facing the world currently. These issues cover an ever-increasing spectrum of toxins, threats, and deterioration of the environment through ever regions. The most powerful things are acidic deposition, loss of stratospheric ozone, and anthropogenic climate change. That greenhouse gases are arguably the most crucial ecological disaster related to energy consumption. Extensive reading amounts of greenhouse gas emissions are growing the way under which these greenhouse gases absorb heat again from earth's surface, thus raising the surface earth's temperature and as a result, rising sea levels. A number of alternative approaches to the emerging pollution issues caused by toxic pollution of contaminants have appeared recently. Even so, green energy seems to be among the most important options. In order to accomplish the power, technical and political advantages that renewable sources provide an interconnected collection of practices, like R&D, technology evaluation, policy implementation, and technological transition, must be carried out as needed. Sustainability requires a consistent future of energy supplies that throughout the longer term is freely and affordably accessible at a fair market price that can be used for all the activities needed despite having harmful social ramifications. Supplies of energy resources such as fossil fuels and uranium are commonly recognized to be unbounded; other sources of energy like solar, wind, and tidal currents are widely known to be stable and thus abundant for a comparatively long period. The use of green energy options and technology is a core component of sustainable growth based on the information: far less effects on the environment, greater stability, lack of implementation, and the prospect of decentralization. Growing the earth's population needs the concept and effective application of sustainable growth [12].

4 Ecological Footprint

Ecological footprint is used as an effective engagement mechanism, has a specific capability in the government sense, is restricted in nature, needs to be more closely associated only with United Nations Ecological and Financial Accountability Framework, and is most valuable as part of a set of metrics. Key problems in the examination of approaches are neither of the main mechanisms found that can resolve all related concerns and challenges simultaneously; centered on bio-productivity estimates on net primary output is a viable candidate; advancements in the relation between bio-productivity and ecological services including ecosystems are being produced by the flexible ecological footprint principle and the HANPP predictor; ecologically input–output review gives a lot of advantages for enhancing ecological footprint computations as well as further variants; including the electricity idea or the participation of additional toxins; are not considered to provide such a major change to the effectiveness of ecological footprint government policy [67].

Just like worries about rising environmental destruction, sustainable development study is becoming increasingly relevant. The ecological footprint index is a budget-planning instrument that is commonly used in the study of sustained development. While ecological footprint is beneficial to other conservation research approaches with a measurable metric, it also has drawbacks on the research of some crucial ecological concerns, such as unsustainable land usage, degradation of renewable resources, and incorrect calculation of carbon emissions, that is the largest element of ecological footprint. The main strengths of the ecological footprint approach make it generally available in the planning strategies of a variety of social or sub-national authorities. Compared to other measurement metrics, ecological footprint is structured to capture and summaries dynamic and huge phenomena in a simplistic and reliable manner. The balance between disadvantages and advantages is important for the use of this metric [71].

Measurements are at the core of the sustainable development, and the ecological footprint method is increasingly becoming an effective way to assess biodiversity. Because it embraces a similar norm, the findings of the EF analysis model may be equivalent and readily interpreted. The aim of the ecological footprint test is to analyze the need of a specific population being studied of existence by either a standard system, hence a bio-productive region, as well as its objective is to approximate the stable condition by examining if the demand exceeds the overall supply. The ecological footprint method has already been extended to assess the survival of so many countries, cities, and territories, and of the planet itself. In addition, its area of application has been expanded to include property demand forecasting, evaluation of travel resilience, environmental assessment, and so on. For the growth of the program, the ecological footprint approach has been steadily changed and revised. In order to obtain integrated trade statistics, some studies used input and output measurement instead of the initial calculation process. The local profitability variable method is used to demonstrate the geographic characteristics. Power to predict the ecological footprint and improve technology growth in the natural, economic, and social fields [31].

5 Renewable Energy and Its Ecological Footprint

Income activity, combined with increasing energy consumption in Asian nations, has been unparalleled in recent years. In another hand, the electricity used in the Asian market is largely non-renewable, which may have consequences for sustained development. Previous research that examined the energy-growth-environmental connection for this area is unreliable in spite of the metrics used to assess environmental efficiency and the measurement methods used. Even so, this analysis examines the effect of renewable and non-renewable energy use, industrial development and urbanization on a more reliable environmental metric (ecological footprint) since 1990, thus regulating trade. First- and Second-generation source unit and co-integration experiments are implemented in the sense of proof of cross-sectional dependency. The results suggest that economic development, commerce, and non-renewable energies lead contributes to global destruction in Asian nations [40].

Large population nations, like India and China, have to boost employment in order to reach an optimal level of profits. Even so, economic activity drives extraction of natural resources; the influence of size on nature and technological results in

China and India is dominant. National governments must concentrate on a much more efficient usage of natural resources in tandem with increasing incomes. The incorporation of green energy at a higher pace in the energy mix is advocated. As a result, with increasing wages, more expenditure can be dedicated to developments in green energy initiatives. In addition, income alone will control emissions, along with certain stringent environmental steps that are needed to encourage further efficient use of resources [38, 63].

In another side, political leaders in the majority of the test countries like Brazil, Russia, and South Africa are continuous integration the share of renewables throughout their electricity sector, handling their environmental available resources, and monitoring their urbanization practices in a similar manner to their existing consequences for a prosperous future. As evidential findings show, sustainable energy reduces ECF; thus with increased economic activity, more expenditure in renewable energy is required, as energy is mainly connected to three dimensions of sustainable development, and green system is the energy option to enhancing financial, economic, and environmental conservation [63, 64].

The long-term balancing partnership confirmed the critical role of green energy use in enhancing the sustainability of the atmosphere and health—as fewer air pollutants increase air pollution. Even so, there is any need to develop and reinforce the environmental management system of the blocks under investigation, if certain Member States are yet to cooperate with the Paris Treaty in order to minimize global pollution. The report also showed that a decrease in environmental emissions in one Member State is unlikely to result in environmental protection for all National Governments. The aim must thus become ecological protection for all Member States without any isolated situations, thus implying a shared path toward meeting the sustainability growth targets by 2030 [1].

Renewable energy does not make a substantial contribution to air sustainability, whereas non-renewable energy use contributes substantially to ecological depletion. One-way cause and effect flows from urbanization, economic development, and energy consumption to environmental destruction. One way to mitigate this harm is for countries to accept and encourage the usage of renewable energy sources [39].

6 Conclusion

This chapter presents about the renewable energy types, importance of renewable energy, ecological footprint, and the relationship between the renewable energy and the ecological footprint. Renewable energies are known as green energy sources as well as the efficient usage of these products minimizes ecological consequences, creates minimal ancillary emissions, and are viable on the basis of present and emerging socioeconomic empowerment. The sun is the center of all energy. Heat and light are the main sources of solar energy. Renewable energy solutions offer an outstanding chance to minimize greenhouse gases and stop global warming by replacing conventional fossil fuels. Furthermore, the analysis suggested that the increased utilization clean energy has substantially decreased the ecological foot-print of the area. Even though a growth of population has resulted in an increase in air pollution in these nations. Similarly, the effect of lifespan on the ecological footprint is favorable but negligible.

References

1. Alola AA, Bekun FV, Sarkodie SA (2019) Dynamic impact of trade policy, economic growth, fertility rate, renewable and non-renewable energy consumption on ecological footprint in Europe. Sci Total Environ 685:702–709
2. Atilgan B, Azapagic A (2015) Life cycle environmental impacts of electricity from fossil fuels in Turkey. J Clean Prod 106:555–564
3. Aydin C, Esen Ö, Aydin R (2019) Is the ecological footprint related to the Kuznets curve a real process or rationalizing the ecological consequences of the affluence? Evidence from PSTR approach. Ecol Ind 98:543–555
4. Barbier E (2002) Geothermal energy technology and current status: an overview. Renew Sustain Energy Rev 6(1–2):3–65
5. Barbir F, Veziroğlu TN, Plass HJ Jr (1990) Environmental damage due to fossil fuels use. Int J Hydrogen Energy 15(10):739–749
6. Bellarmine GT, Urquhart J (1996) Wind energy for the 1990s and beyond. Energy Convers Manage 37(12):1741–1752
7. Bello MO, Solarin SA, Yen YY (2018) The impact of electricity consumption on CO_2 emission, carbon footprint, water footprint and ecological footprint: the role of hydropower in an emerging economy. J Environ Manage 219:218–230
8. Biermann E, Grupp M, Palmer R (1999) Solar cooker acceptance in South Africa: results of a comparative field-test. Sol Energy 66(6):401–407
9. Bilgen S, Kaygusuz K, Sari A (2004) Renewable energy for a clean and sustainable future. Energy Sources 26(12):1119–1129
10. Chiari L, Zecca A (2011) Constraints of fossil fuels depletion on global warming projections. Energy Policy 39(9):5026–5034
11. Demirbas A (2000) Recent advances in biomass conversion technologies. Energy Edu Sci Technol 19–40
12. Dincer I (2000) Renewable energy and sustainable development: a crucial review. Renew Sustain Energy Rev 4(2):157–175
13. Dincer I (2001) Environmental issues: Ii-potential solutions. Energy Sources 23(1):83–92
14. Dögl C, Holtbrügge D (2000) Competitive advantage of German renewable energy firms in Russia-an empirical study based on Porter's diamond. J East European Manage Stud 34–58
15. Edenhofer O, Pichs-Madruga R, Sokona Y, Seyboth K, Kadner S, Zwickel T, Eickemeier P, Hansen G, Schlömer S, von Stechow C, Matschoss P (2011) (eds) Renewable energy sources and climate change mitigation: special report of the intergovernmental panel on climate change. Cambridge University Press
16. Ellabban O, Abu-Rub H, Blaabjerg F (2014) Renewable energy resources: current status, future prospects and their enabling technology. Renew Sustain Energy Rev 39:748–764
17. Eltamaly AM (2013) Design and implementation of wind energy system in Saudi Arabia. Renew Energy 60:42–52
18. Farhad S, Saffar-Avval M, Younessi-Sinaki M (2008) Efficient design of feedwater heaters network in steam power plants using pinch technology and exergy analysis. Int J Energy Res 32(1):11
19. Gonzalo AP, Marugán AP, Márquez FP (2019) A review of the application performances of concentrated solar power systems. Appl Energy 255:113893

20. Goswami DY, Vijayaraghavan S, Lu S, Tamm G (2004) New and emerging developments in solar energy. Sol Energy 76(1–3):33–43
21. Hall DO (1991) Cooling the greenhouse with bioenergy. Nature 353(6339):11–12
22. Hao XH, Guo LJ, Mao X, Zhang XM, Chen XJ (2003) Hydrogen production from glucose used as a model compound of biomass gasified in supercritical water. Int J Hydrogen Energy 28(1):55–64
23. Herbert GJ, Iniyan S, Sreevalsan E, Rajapandian S (2007) A review of wind energy technologies. Renew Sustain Energy Rev 11(6):1117–1145
24. Herzog AV, Lipman TE, Kammen DM (2001) Renewable energy sources. Encyclopedia of life support systems (EOLSS). Forerunner Volume-'Perspectives and overview of life support systems and sustainable development 76
25. Holm-Nielsen JB, Al Seadi T, Oleskowicz-Popiel P (2009) The future of anaerobic digestion and biogas utilization. Biores Technol 100(22):5478–5484
26. Höök M, Tang X (2013) Depletion of fossil fuels and anthropogenic climate change—a review. Energy Policy 52:797–809
27. Hossain AS, Salleh A, Boyce AN, Chowdhury P, Naqiuddin M (2008) Biodiesel fuel production from algae as renewable energy. Am J Biochem Biotechnol 4(3):250–254
28. Islam MR, Mekhilef S, Saidur R (2013) Progress and recent trends of wind energy technology. Renew Sustain Energy Rev 21:456–468
29. Jäger-Waldau A (2020) Snapshot of photovoltaics—february 2020. Energies 13(4):930
30. Jensen SH, Larsen PH, Mogensen M (2007) Hydrogen and synthetic fuel production from renewable energy sources. Int J Hydrogen Energy 32(15):3253–3257
31. Jiang Y, Wang Y, Pu X, Wang J (2005) Review and prospect of the application of ecological footprint model. Prog Geogr 2
32. Kaygusuz K (2009) Wind power for a clean and sustainable energy future. Energy Sources Part B 4(1):122–133
33. Kothari R, Tyagi VV, Pathak A (2010) Waste-to-energy: a way from renewable energy sources to sustainable development. Renew Sustain Energy Rev 14(9):3164–3170
34. Kumar A, Kandpal TC (2007) CO_2 emissions mitigation potential of some renewable energy technologies in India. Energy Sources Part A 29(13):1203–1214
35. Lee AH, Chen HH, Chen S (2015) Suitable organization forms for knowledge management to attain sustainable competitive advantage in the renewable energy industry. Energy 89:1057–1064
36. Lotfalipour MR, Falahi MA, Ashena M (2010) Economic growth, CO_2 emissions, and fossil fuels consumption in Iran. Energy 35(12):5115–5120
37. Martins F, Felgueiras C, Smitkova M, Caetano N (2019) Analysis of fossil fuel energy consumption and environmental impacts in European countries. Energies 12(6):964
38. Munasinghe M (2002) The sustainomics trans-disciplinary meta-framework for making development more sustainable: applications to energy issues. Int J Sustain Dev 5(1–2):125–182
39. Nathaniel S, Anyanwu O, Shah M (2020) Renewable energy, urbanization, and ecological footprint in the Middle East and North Africa region. Environ Sci Pollut Res 11:1–3
40. Nathaniel S, Khan SA (2020) The nexus between urbanization, renewable energy, trade, and ecological footprint in ASEAN countries. J Clean Prod 272:122709
41. Omara AA, Abuelnuor AA, Mohammed HA, Habibi D, Younis O (2020) Improving solar cooker performance using phase change materials: a comprehensive review. Sol Energy 207:539–563
42. Owusu PA, Asumadu-Sarkodie S (2016) A review of renewable energy sources, sustainability issues and climate change mitigation. Cogent Eng 3(1):1167990
43. Panayotou T (1993) Empirical tests and policy analysis of environmental degradation at different stages of economic development. International Labour Organization
44. Panwar NL, Kaushik SC, Kothari S (2011) Role of renewable energy sources in environmental protection: A review. Renew Sustain Energy Rev 15(3):1513–1524
45. Parida B, Iniyan S, Goic R (2011) A review of solar photovoltaic technologies. Renew Sustain Energy Rev 15(3):1625–1636

46. Rabaia MK, Abdelkareem MA, Sayed ET, Elsaid K, Chae KJ, Wilberforce T, Olabi AG (2020) Environmental impacts of solar energy systems: a review. Sci Total Environ 754:141989
47. Ram JP, Manghani H, Pillai DS, Babu TS, Miyatake M, Rajasekar N (2018) Analysis on solar PV emulators: a review. Renew Sustain Energy Rev 81:149–160
48. Razykov TM, Ferekides CS, Morel D, Stefanakos E, Ullal HS, Upadhyaya HM (2011) Solar photovoltaic electricity: current status and future prospects. Sol Energy 85(8):1580–1608
49. Rybach L (2003) Geothermal energy: sustainability and the environment. Geothermics 32(4–6):463–470
50. Rybach L, Mégel T, Eugster WJ (2000) At what time scale are geothermal resources renewable. In: Proc. World Geothermal Congress 2000, vol 2, pp 867–873
51. Şahin AD (2004) Progress and recent trends in wind energy. Prog Energy Combust Sci 30(5):501–543
52. Shepovalova OV (2015) Energy saving, implementation of solar energy and other renewable energy sources for energy supply in rural areas of Russia. Energy Procedia 4:1551–1560
53. Sheth S, Shahidehpour M (2004) Geothermal energy in power systems. In: IEEE power engineering society general meeting. IEEE, pp 1972–1977
54. Sims RE (2003) Bioenergy to mitigate for climate change and meet the needs of society, the economy and the environment. Mitig Adapt Strat Glob Change 8(4):349–370
55. Singh R, Leslie JD (1980) Economic requirements for new materials for solar photovoltaic cells. Sol Energy 24(6):589–592
56. Singh S, Bhatti TS, Kothari DP (2004) Indian scenario of wind energy: problems and solutions. Energy Sources 26(9):811–819
57. Solarin SA, Tiwari AK, Bello MO (2020) A multi-country convergence analysis of ecological footprint and its components. Sustain Cities Soc 46:101422
58. Sreekrishnan TR, Kohli S, Rana V (2004) Enhancement of biogas production from solid substrates using different techniques—a review. Biores Technol 95(1):1
59. Srirangan K, Akawi L, Moo-Young M, Chou CP (2012) Towards sustainable production of clean energy carriers from biomass resources. Appl Energy 100:172–186
60. Tester JW, Anderson BJ, Batchelor AS, Blackwell DD, DiPippo R, Drake EM, Garnish J, Livesay B, Moore MC, Nichols K, Petty S (2006) The future of geothermal energy. Massachusetts Inst Technol 26:358
61. Topcu YI, Ulengin F (2004) Energy for the future: an integrated decision aid for the case of Turkey. Energy 29(1):137–154
62. Turner J, Sverdrup G, Mann MK, Maness PC, Kroposki B, Ghirardi M, Evans RJ, Blake D (2008) Renewable hydrogen production. Int J Energy Res 32(5):379–407
63. Ulucak R, Khan SU (2020) Determinants of the ecological footprint: role of renewable energy, natural resources, and urbanization. Sustain Cities Soc 54:101996
64. Ulucak R, Yücel AG, Koçak E (2019) The process of sustainability: from past to present. In: Environmental Kuznets Curve (EKC). Academic Press, pp 37–53
65. Vengadesan E, Senthil R (2020) A review on recent development of thermal performance enhancement methods of flat plate solar water heater. Sol Energy 206:935–961
66. Weiland P (2010) Biogas production: current state and perspectives. Appl Microbiol Biotechnol 85(4):849–860
67. Wiedmann T, Barrett J (2010) A review of the ecological footprint indicator—perceptions and methods. Sustainability 2(6):1645–1693
68. Wilberforce T, Baroutaji A, El Hassan Z, Thompson J, Soudan B, Olabi AG (2019) Prospects and challenges of concentrated solar photovoltaics and enhanced geothermal energy technologies. Sci Total Environ 659:851–861
69. Youm IS, Sarr J, Sall M, Kane MM (2000) Renewable energy activities in Senegal: a review. Renew Sustain Energy Rev 4(1):75–89
70. Zakhidov RA (2008) Central Asian countries energy system and role of renewable energy sources. Appl Solar Energy 44(3):218–223
71. Zhang L, Dzakpasu M, Chen R, Wang XC (2017) Validity and utility of ecological footprint accounting: a state-of-the-art review. Sustain Cities Soc 32:411–416

72. Zhou L, Schwede DB, Appel KW, Mangiante MJ, Wong DC, Napelenok SL, Whung PY, Zhang B (2019) The impact of air pollutant deposition on solar energy system efficiency: an approach to estimate PV soiling effects with the Community Multiscale Air Quality (CMAQ) model. Sci Total Environ 651:456–465
73. Züttel A, Remhof A, Borgschulte A, Friedrichs O (2010) Hydrogen: the future energy carrier. Philos Trans R Soc A: Math Phys Eng Sci 368(1923):3329–3342

Pakistan Ecological Footprint and Major Driving Forces, Could Foreign Direct Investment and Agriculture Be Among?

Edmund Ntom Udemba

Abstract Pakistan economy is characterized with good economic performance with agriculture and foreign direct investment (FDI) as pillars of the economy but remains among top emission producing countries. This calls for investigation of the economy with respect to its contribution toward climate change. Many indicators have been proposed as the best measures of environment performance but less consideration to ecological footprint as the right measure of environment, this current study adopts ecological footprint as a proxy to environment performance. FDI and agriculture are equally selected as among the variables in this study. Pakistan country specific data from 1970 to 2018 was applied to this research for clear insight both in short- and long-run investigation. ARDL bound and granger causality tests were applied. FDI and agriculture were found impacting ecological footprint negatively. Causal transmission was found passing to ecological footprint from FDI. Nexus was found among the selected variables pointing toward negative impact on ecological footprint. Hence, pollution haven hypothesis was confirmed for Pakistan.

Keywords Economic growth · Energy use · FDI · Agriculture · PHH · Pakistan

1 Introduction

Ecological footprint as a measure of environmental degradation has been largely overlooked mostly by the developing nations. Few scholars have adopted ecological footprint as a measure to study environmental performance with little emphasis on the meaning and correlation of ecological footprint and the quality performance of environment. Misconception of the indicator has made some of the scholars to posit ecological footprint as responsible for environmental deterioration. For clear insight on the linkage of the ecological footprint with environment performance, it is necessary to consider the meaning of both over shooting of the Earth and ecological deficit.

E. N. Udemba (✉)
Faculty of Economics Administrative and Social Sciences, Istanbul Gelisim University, Istanbul, Turkey
e-mail: eudemba@gelisim.edu.tr; eddy.ntom@gmail.com

© The Author(s), under exclusive license to Springer Nature Singapore Pte Ltd. 2021   109
S. S. Muthu (ed.), *Assessment of Ecological Footprints*,
Environmental Footprints and Eco-design of Products and Processes,
https://doi.org/10.1007/978-981-16-0096-8_5

According to Global Network Footprint (GNF), ecological footprint defined as the utilization of landed and water resources for production of all resources consumed by the population. From the definition of ecological footprint, over utilization of natural resources than the available or regenerated resources by the ecosystem is known as ecological deficit, and when the available space occupied by humanity and its impact is greater than the ecosystem, it is considered as Earth's overshoot [26]. The presence of both ecological deficit and earth's overshoot is what will prompt environment degradation. Also, the mitigation of environment through ecological footprint can be viewed from the demand and supply on and from the ecosystem. The demand on ecosystem is seen from the human activities on soil especially for farming purposes and also demands for space for city or urban infrastructure, and forest to absorb its carbon dioxide emissions from fossil fuels. The supply of ecosystem represents the available land and sea area which includes lands for forestry, grazing, cropping, fishing grounds, and built-up land. Ecological footprint (EFP) can be measured and calculated for space occupied by individual, city, region, nation, country, and the complete planet. EFP of a person or a country is determined by how much of biocapacity used and how efficiency this is being produced.

This study is centered on mitigating Pakistan's ecological footprint with agriculture and foreign direct investment. Pakistan is among the vulnerable countries to climate change through low environmental performance Bhandari [7]. The economic profile of Pakistan has positioned agriculture and foreign direct investment (FDI) as among the sustenance of the country's economy with high reliance on imported gas and oil as the main energy source and consumption Nabi [25]. Agriculture gives one-fifth of the GDP and two-fifth of employment in Pakistan Bergan [6]. Among the agricultural products of the country are cotton, wheat, rice, sugarcane, fruits, vegetables, milk, beef, mutton, and eggs. Pakistan government has embarked on some policies to boost the country's attractiveness for investors, such as tax incentives for floating up of industries in sectors like energy, ports, highways, and software. Also, export-processing zones are created by government to encourage the foreign investment. This involves exemption from all form of taxes and duties on equipment, machinery and materials, and access to Export Processing Zone Authority one window services [43]. Recently, FDI inflows to Pakistan have shown a positive and increasing pattern from US $1.7 billion in 2018 to US $2.2 and US $34.8 billion, respectively, in 2019 [45], UNCTAD's 2020 World Investment Report). Countries that top most in Pakistan's investment are China, United Kingdom (UK), South Korea, and Japan. The primary recipients of the FDI in Pakistan are financial sector, chemical industry, and construction [45], UNCTAD's 2020 World Investment Report). Pakistan's vulnerability to climate change is not unseparated from the activities emanating from the activities from energy, agricultural and industrial cum investment sectors. With ranking of 8th in the Germanwatch long-term climate risk index, Pakistan is recorded as among the world's most vulnerable countries to climate change [17]. As recorded Chaudhry [11], the country's share of fossil fuel energy sources consumption has risen to almost 63% currently. Also, the need to boost the industrial production which equally contains the investors activities has prompted heavy reliance on imported fossil fuels and boosting of coal power plants which would further compound its vulnerability to climate changes. The dirty practices that exist in execution of farming practices such as application of chemicals both in crop planting and fisheries, planting of rice in riverine and coastal areas deforestation for constructions, herders' activities and others poses environmental danger to the sustainable development with regard to environmental impact.

Following the domineering of agriculture in the country's profile of economic operation and the loose policies to encourage investors, especially foreigners with over reliance on fossil fuels as source of energy consumption, author seeks to investigate the vulnerability level of Pakistan to climate changes. Ecological footprint is considered the correct indicator to measure the environment performance because of the complexity of the sectors involved in this study. Few scholars have utilized ecological footprint in their studies as a comprehensive measure of environmental dilapidation and performance [3, 5, 9, 15, 19, 24, 28, 33, 37, 39, 40, 42]. This is not the first study that has tried to investigate the environment performance with ecological footprint, the only difference is in the choice of variable. Scholars have used particular variables such as economic growth [27], globalization [33], and social political factors [10] and tourism [22] to study the ecology. However, this study is different from other works with the consideration of agricultural sector of Pakistan which is considered as among the sustainable sectors of the country with much practices that exposes the environment to degradation. Also, FDI is considered among the tentative sectors to induce environmental dilapidation of Pakistan considering the loose hand in checkmating the excesses of the sector from the government policies. Diverse scientific approaches such as Autoregressive Distributed Lag (ARDL)-bound testing, granger causality, and some diagnostic tests are utilized in this study to ensure accuracy in the findings.

Previous studies have tried to investigate driving forces of ecological footprint such as Charfeddine and Ben Khediri found that trade, economic growth, and urbanization impact negatively on the quality of environment: Kasman and Selman [21] found economic growth and trade important forces in determining the quality of environment and [27] found environment Kuznets curve for the countries for the economic growth force, Solarin and Al-mulali, 2018 and Udemba et al. [38] for foreign direct investment [22] for tourism force, and [33] for globalization.

The rest sections of this study are the theoretical background, data and methodology, results and empirical analyses, and conclusion.

2 Theoretical Background

The hypothetical background of this work is anchored on environmental Kuznets curve (EKC) and pollution haven hypothesis (PHH). Environmental Kuznets curve hence taken as EKC is first conceived and applied by Simeon Kuznets in his study of income inequality between rural farmers and city employees. He found a turning point that assures the closing of the income gap. Since then, some energy and environmental economists [20, 29] have adopted and applied same hypothesis in studying the interaction between economic growth and environment performance. They hypothesized that even though the economic growth will cause environmental dilapidation, the scenario will eventually change to the betterment of the environment. This they argued based on the awareness and increased sensitivity of people at the later stage of the economic growth on the importance of maintaining good quality of environment. For the effectiveness of this study based on EKC, squared GDP per capita is employed to monitor the economic growth pattern. Pollution haven hypothesis hence taken as PHH is among the hypothetical backgrounds of this study. This hypothesis is based on the implication of FDI and the activities of foreign investors on the environmental performance of the host country. There have been diverse views on this hypothesis, with some scholars [12, 18, 37, 39] for India and Turkey, and [13, 34, 35] favoring the negative implication of FDI on the environment quality, while others [1, 4, 8, 23, 38, 44] for Indonesia) are favoring the positivity of FDI on the both economy and environment. The negative influence of FDI is perceived when the activities of the foreign investors induce environment dilapidation through dirty practice of production with fossil fuel energy sources. However, the positive impact of FDI is perceived when the foreign investors are impacting favorable to the growth of economy through spillover or leakage effects such as employment expansion, technical and skill transfer and economics of scale, and equally impacting the environment quality through improved technology and innovation that enhances clean production void of pollution. Pollution haven hypothesis (PHH) is valid when FDI is impacting unfavorable on the environment, and not valid when FDI is favorable and this is called pollution halo hypothesis (PHH). When investment in a country introduced a dirty production with high utilization of fossil fuels that violates environmental regulations it supports pollution haven hypothesis (PHH). Contrary to the negative impact of foreign investment, pollution halo hypothesis is established if the positive impact of foreign direct investment such as technical transfer, job creation, and economics of skill outweighs its negative impact.

3   Data, Methodology, and Modeling

Pakistan country-specific data from 1970 to 2018 was applied to this research for clear insight both in short- and long-run investigation. The data was made up of different variables sourced from different source. Per capita ecological footprint, per capita GDP (constant 2010 US$), squared per capita GDP (constant 2010 US$), agriculture (agriculture, forestry, and fishing value added constant 2010 US$), foreign direct investment (% of GDP), energy use (million tonnes oil equivalent). Apart from ecological footprint and energy use that were garnered from Global Network of Footprint [26] and British Petroleum (BP) statistical review [32], respectively, all other variables were sourced from World Development Index (WDI) [41].

Methodologies applied in this study are descriptive statistics for ascertaining the normality of the data distribution, test of stationarity for tracing of the pattern of the data and order of integration of the variables. ARDL-bound testing was equally utilized for cointegration testing in determination of the possibility of long-run relationship among the variables. Diagnostic tests such as test for serial correlation and heteroscedasticity, cumulative sum square (CUSUM and CUSUM^2) were all utilized in robust checking of the findings from other approaches. Also, granger causality (VECM block exogeneity) was equally adopted in complementing the linear and nonlinear relationship that exist among the variables in both short run and long run. It (granger causality) was specifically applied to ascertain if there the variables and their lags establish forecasting ability among themselves.

Modeling of this study is anchored on the hypothetical backup of this work (EKC) and ARDL-bound testing according to Pesaran et al. [31] model specification of cointegration. Model specification is expressed in four (4) different levels; at equation level, at econometric model and in empirical model and in ARDL cointegration model as follows:

Equation form

$$EFP = (GDP, GDP^2, FDI, AGR, EU) \tag{1}$$

Econometric form

$$EFP = b_1 + b_2 GDI_{it} + b_3 GDP_{it}^2 + b_4 FDI_{it} + b_5 AGR_{it} + b_6 EU_{it} + \mu_{it} \tag{2}$$

Empirical form

$$\ln EFP = b_1 + b_2 \ln GDP_{it} + b_3 \ln GDP_{it}^2 + b_4 FDI_{it} + b_5 \ln AGR_{it} + b_6 \ln EU_{it} + \mu_{it} \tag{3}$$

Error correction model

$$\ln EFP_t = \partial + b_1 \ln EFP_{t-1} + b_2 \ln GDP_{t-1} + b_3 \ln GDP_{t-1}^2 + b_4 FDI_{t-1}$$
$$+ b_5 \ln EU_{t-1} + b_6 \ln AG_{t-1} + \sum_{i=0}^{p-1} a_1 \Delta \ln EFP_{t-i} + \sum_{i=0}^{q-1} a_2 \Delta \ln GDP_{t-i}$$
$$+ \sum_{i=0}^{q-1} a_3 \Delta \ln GDP_{t-i}^2 + \sum_{i=0}^{q-1} a_4 \Delta FDI_{t-i} + \sum_{i=0}^{q-1} a_5 \Delta \ln EU_{t-i}$$
$$+ \sum_{i=0}^{q-1} a_6 \Delta \ln AG_{t-i} + ECM_{t-i} + \mu_t$$

$$\sum_{i=0}^{q-1} a_2 \Delta \ln GDP_{t-i} + \sum_{i=0}^{q-1} a_3 \Delta \ln GDP_{t-i}^2 + \sum_{i=0}^{q-1} a_4 \Delta FDI_{t-i} + \sum_{i=0}^{q-1} a_5 \Delta \ln EU_{t-i}$$
$$+ \sum_{i=0}^{q-1} a_6 \Delta \ln AG_{t-i} + ECM_{t-i} + \mu_t \tag{4}$$

From Eqs. 1 to 4 EFP, GDP, GDP^2, FDI, AGR, and EU represent per capita ecological footprint, per capita GDP and squared per capita GDP, foreign direct investment, agriculture and energy use, respectively. Ecological footprint is a proxy to measure the environmental dilapidation, GDP and GDP^2 are measures of economic growth which are utilized for the sake of analyzing EKC hypothesis. Where t and μ_{it} represent time and error term, b_0 & a_0 represent the long-run and short-run coefficients to be estimated. From Eq. (4) Δ and ECM$_{t-i}$ denote sign of first difference and error correction model. To test for cointegration, it is hypothesized that there is no cointegration with null hypothesis as following:

Null hypothesis ($H_0 = b_16 = 0$) against alternative hypothesis ($H_1 = b_16 \neq 0$). The cointegration is identified with comparing F-stat with the critical values of upper and lower bounds in bound tests. If the F-stat is greater than the critical values at 1.5 or 10% significant level, cointegration is established and vice versa.

4 Empirical Result and Explanations

4.1 Descriptive Statistics

Descriptive analysis of the data was obtained in order to inspect the normality and stability of data applied in this study. With the outcomes of Kurtosis and Jarque-Bera, some level of normality was confirmed. Hence, all the statistics under kurtosis were below 3, and the probability of the Jarque-Bera outcome was all insignificant except for the value of FDI which is insignificant in affecting the normality of the entire data. Considering the variability of the variables in use, GDP is the most varied indices followed by energy use. With the normality of the data, linear analysis will be effective in this study. The output is shown in Table 1.

4.2 Stationarity Test

Stationarity of the data and order of integration of the variables were confirmed with unit root test. Time series estimations are known with unstable variables because of the likelihood of structural changes that may obstruct the stability of the trend in economic variables. For effective analysis of these estimations, two approaches were applied, hence, augmented Dickey–Fuller [16] and perron [30]. This will permit for robust check among the two methods. Unit root was confirmed with mixed order of integration among the variables. The unit root output paved way for the decision on the appropriate method to apply further in testing the cointegration. The unit root output is displayed on Table 2.

Table 1 Descriptive statistics

	EFP	GDP	EU	FDI	AGRIC
Mean	0.757059	754.7259	29.59761	0.746787	2.59E+10
Median	0.765958	787.3003	26.68124	0.566385	2.41E+10
Maximum	0.926129	1117.518	66.85071	3.668323	4.69E+10
Minimum	0.633942	450.3759	6.534027	-0.062662	1.09E+10
Std.Dev.	0.088139	198.0054	18.79940	0.798594	1.19E+10
Skewness	0.072895	-0.000256	0.391197	2.217013	0.316721
Kurtosis	1.707545	1.851147	1.764700	7.890248	1.713076
Jarque-Bera	3.312902	2.584731	4.187126	85.33461	4.029116
Probability	0.190815	0.274620	0.123247	0.000000	0.133379
Sum	35.58179	35472.12	1391.088	35.09898	1.22E+12
SumSq. Dev.	0.357348	1803483.	16257.20	29.33660	6.49E+21
Observations	47	47	47	47	47

Source Authors computation

Table 2 Unit root test

Variables		@ LEVEL		First Diff		
	With intercept	Intercept & trend	With intercept	Intercept & trend	Decision	
PP						
LEFP	−1.1265	−2.5289	−8.1305***	−8.0683***	I(1)	
LGDP	1.3567	−2.1622	−3.9969***	−4.1957***	I(1)	
LEU	2.5669	−1.7461	−3.4438**	−4.0284**	I(1)	
FDI	−1.9540	−2.0496	−4.4422***	−4.4065***	I(1)	
LAGR	2.8785	−5.6037***	−7.7651***	−9.2561***	MIXED	
ADF						
LEFP	−1.2248	−2.4877**	−8.1377***	−8.0729***	MIXED	
LGDP	0.8953	−1.9474	−4.0830***	−4.2549***	I(1)	
LEU	1.3510	−2.0178	−3.4438***	−3.9700***	I(1)	
FDI	−3.0025**	−3.3474*	−4.7110***	−4.6877***	MIXED	
LAGR	2.0462	−2.7743	−7.7578***	−8.7232***	I(1)	

Notes (a): (*) Significant at the 10%; (**) Significant at the 5%; (***) Significant at the 1% (b): *P*-value according to (1) Maclean et al. one-sided *P*-values
Source Author's computation

4.3 Linear and Cointegration (Short and Long Run)

For the linear and cointegration estimations, ARDL-bound test was utilized for accurate analysis. From the output of unit root test, a mixed order of integration is recorded which is unsuitable for application of any other approach of cointegration. In this case, ARDL is considered the best approach for estimating the cointegration and by extension the short- and long-run linear analysis. The output is displayed in Table 3. Optimal lag was estimated with Akaike Information Criteria (AIC) and the output is 2. Cointegration was estimated with bound test and null hypothesis was rejected

Table 3 ARDL dynamics and cointegration test of EFP model

Variables	Coefficients	SE	t-statistics	P-value
Short-path				
D(LGDP)	−0.005523	0.000778	−7.095405	0.0000***
D(LGDP^2)	3.31E−06	5.04E−07	6.554116	0.0000***
D(FDI)	0.021838	0.007271	3.003627	0.0056***
D(LEU)	0.008030	0.002967	2.706143	0.0115***
D(AGR)	1.15E−11	4.73E−12	2.426328	0.0219**
CointEq (−1)*	−0.780949	0.089120	−8.762906	0.0000***
Long-path				
LGDP	−0.005523	0.001457	−3.791664	0.0007***
LGDP^2	3.31E−06	9.89E−07	3.343400	0.0024***
FDI	0.021838	0.010250	2.130443	0.0421**
LEU	0.008030	0.004000	2.007620	0.0544**
LAGR	1.15E−11	4.73E−12	2.426328	0.0219***
Constant	0.302741	0.079667	3.800073	0.0007***
R^2	0.972668			
Adj.R^2	0.958026			
D.Watson	1.639145			
Bound test (Long-path)				
F-statistics	9.033945***	$K = 5$,@ 1%	I(0)bound = 3.06	I(1)bound = 4.15
Wald test (short-path)				
F-statistics	66.43016			
P-value	0.000000			
Serial Correlation test				
F-statistics	1.349574			
Chi-square	6.132587			
P-value	0.2810			
Heteroscedasticity Test				
F-statistics	0.553412			
Chi-square	10.06171			
P-value	0.8852			

Note *, **, *** Denotes rejection of the null hypothesis at the 1%, 5%, and 10%
Sources Authors computation

at 1 percent significant level affirming the presence of cointegration which attests to possibility of long-run relationship among the variables. Goodness of fit analysis is represented with $R^2 = 0.972668$ and adjusted $R^2 = 0.958026$ which shows the extend (95.8%) of dependent variable that is explained by the explanatory variables, the remaining part (4.2%) is explained by the residuals. Absence of autocorrelation problem from the model and estimation was confirmed with the Durbin Watson and LM tests. While the value of Durbin Watson is 1.639145, the value of LM test shows insignificant in deviance from the null hypothesis of presence of serial correlation. Also, normal distribution of the error term in the model and stability of the data were revealed with test of heteroscedasticity and cumulative sum of square (CUSUM and $CUSM^2$), respectively. The speed of adjustment was confirmed with the error correction model (ECM) at 1% significant level with negative output (-0.780949). This shows the possibility of establishing equilibrium in the long run after deviation at 8% speed of adjustment. The short- and long-run outputs can be interpreted and explained as follow: a negative and significant relationship is found between economic growth and ecological footprint in the first stage of the economic growth but later a positive and significant relationship is established between squared GDP and ecological footprint. This trend is repeated both in short and long run. The outcome projects the initial recovery of environment due to economic growth but the continuous increase in economic growth certainly paved way for sever damage on the quality of environment. This outcome faults the philosophy of standard EKC which believes that continuous economy growth will better the environment. Finding did not support inverted U-shaped of EKC for Pakistan rather N-shaped EKC. This is consistent with the findings by Ahmed and Long [2]. Numerically, one percent increase GDP and GDP^2 will decrease and increase environment degradation by -0.005523 and 0.000000331%, respectively. It is worthy of note that value of the GDP^2 is almost insignificant. This may probably mean that negative effect of economic growth on environment in the long run is so minute. FDI, energy use, and agriculture have positive and significant relationship with ecological footprint, respectively. This shows that Pakistan's environment is impacted negatively by FDI, energy use, and agriculture in the order of 0.021838, 0.008030, and 0.0000000000115%, respectively. The negative impact of agriculture is negligible as reflected in its coefficient value. With the recent quest for preservation of Pakistan's economic growth by the country's authority, it is clear from these findings that economic growth is encouraged at the expense of environment quality. Likewise, the attractiveness of the foreign investors by the Pakistan's authority through relaxed regulations and policy tends to favor only economic growth and pervade the quality expectation of Pakistan's environment. This is in line with findings by [12, 18, 37, 39] for India and Turkey, and Copeland [14, 34, 35] which support pollution haven hypothesis.

4.4 Diagnostic Tests (Cumulative Sum and Cumulative Sum Square, CUSUM and CUSM²)

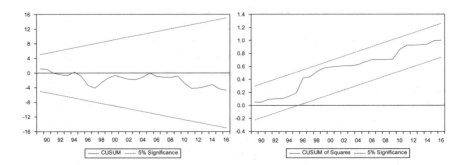

5 Granger Causality Analysis (VECM)

Author applied causality analysis to test forecasting power of the selected variables which will also act as robust check to the findings of short- and long-run cointegration estimations. The linear and quadratic estimations of the ARDL might not give detail insight on the relationships that exist among the variables. Positive and negative interactions of the selected variables are possible without ascertaining the inducer and the direction of the relationship among the variables if granger causality is not considered. For a valid interpretation of the relationship that existed among the variables with forecasting ability, the current study has undertaken the granger causality analysis. Because of the mixed order of integration among the variables, VAR granger causality/block exogeneity was applied, and the outcome is shown in Table 4. From the result, one-way direction causality is found transmitting from energy use to economic growth, from economic growth to FDI and from economic growth to agriculture. Also, two-way causal transmission is found between FDI and ecological footprint, between agriculture and energy use, and between agriculture

Table 4 VECM granger causality analysis/block exogeneity wald tests

Variables					
	ΔLEFP	ΔLGDP	ΔFDI	ΔLEU	ΔLAGR
ΔLEFP	√√ √√	3.299 [0.19]	**5.123 [0.08]**	1.265 [0.53]	4.094 [0.53]
ΔLGDP	3.233 [0.19]	√√ √√	**8.544 [0.01]**	0.319 [0.85]	**6.225 [0.05]**
ΔFDI	**6.336 [0.04]**	2.884 [0.24]	√√ √√	1.532 [0.46]	**5.866 [0.05]**
ΔLEU	0.605 [0.74]	**6.069 [0.05]**	4.507 [0.11]	√√ √√	**8.825 [0.01]**
ΔLAGR	1.337 [0.51]	0.959 [0.61]	**6.062 [0.05]**	**9.011 [0.01]**	√√ √√

Note **Bolden figures** in brackets are the prob. that represent 10%, 5%, and 1% significance resp. while the figures before the brackets are the Chi-squares (χ^2) = Chi-squares (χ^2) [p-values]

and FDI. The granger causality result has shown the linkages among the selected variables. From the result, it is easily deduced that Pakistan economy is energy induced economy. Hence, energy is found transmitting to economic growth. It is worth to know that agriculture and FDI have shown great impact to economic growth, and have established nexus among economic growth, agriculture and FDI. Again, the causal result has complimented the linear result on the relationship between FDI and environment and certified implication of FDI on environment.

6 Conclusion

This research study is based on fact finding on the major drivers of the Pakistan's ecological footprint with an eye on the impact of FDI and agriculture. Author's expectation is derived from the question embedded on the title of this study which asked if FDI and agriculture are among the drivers of Pakistan's ecological footprint. Hence, FDI and agriculture are anticipated to increase ecological footprint there impacting negatively on environment. Scientific approaches such as stationarity test, linear and quadratic version of ARDL and bound testing for cointegration were applied for clear and insightful analyses and findings. N-shaped EKC is established for Pakistan, positive relationship was established between FDI, agriculture, energy use, and ecological footprint. From the granger causality result, a nexus is established among the selected variables exposing Pakistan's economy as energy-led growth. Pollution haven hypothesis was confirmed from both ARDL dynamic linear analysis and granger causal analysis through the involvement of FDI in determining the environment quality. Agriculture and energy use through their impacts on economic growth are strong drivers of Pakistan's ecological footprint. Hence, the question raised in the title of the study is answered with the findings of both analyses (ARDL and granger causality).

Policies targeting balancing the activities of foreign investors in Pakistan's economic pursuit is essential for achievement of sustainable development goals. With the finding of N-shaped EKC, the continuous economic growth is not a criterion for attaining a recoverable state of environment, therefore, policies should be framed around conservative and less carbon energy sources. Long-term goal and model should be formulated to boost the availability of renewable energies such as hydro powers, solar, and oceanic powers.

Conclusively, this study will serve as a reference piece for policy recommendation to the neighboring countries. Further research on this topic is encouraged especially with divergent scientific approaches and other variables such as natural resources and biocapacity and ecological deficit.

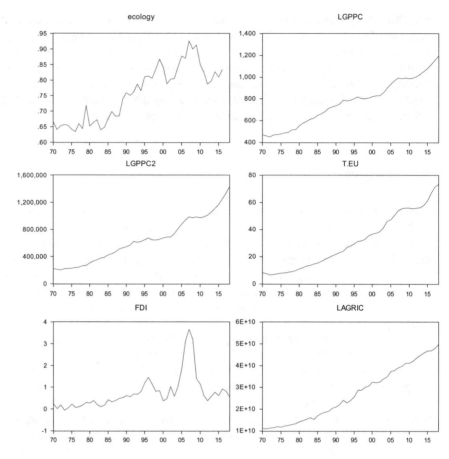

Funding Author wishes to inform the Editor/Journal that no form of funding was received for this research.

Compliance with Ethical Standards

Author wishes to inform the Editor/Journal that there are no conflicts of interest at any level of this study.

References

1. Ahmad N, Du L, Lu J, Wang J, Li HZ, Hashmi MZ (2017) Modelling the CO_2 emissions and economic growth in Croatia: is there any environmental Kuznets curve? Energy 123:164–172
2. Ahmed K, Long W (2013) An empirical analysis of CO_2 emission in Pakistan using EKC hypothesis. J Int Trade Law Policy
3. Al-Mulali U, Weng-Wai C, Sheau-Ting L, Mohammed AH (2015) Investigating the environmental Kuznets curve (EKC) hypothesis by utilizing the ecological footprint as an indicator of environmental degradation. Ecol Ind 48:315–323

4. Alfaro L (2015) Foreign direct investment: effects, complementarities, and promotion. Partners creditors 21–76
5. Barkat K, Mrabet Z, Alsamara M (2016) Does official development assistance for health from developed countries displace government health expenditure in sub-Saharan countries. Econ Bull 36(3):1616–1635
6. Bergan A (1967) Personal income distribution and personal savings in Pakistan: 1963/64. Pakistan Dev Rev 7(2):160–212
7. Bhandari MP (2013) Environmental performance and vulnerability to climate change: a case study of India, Nepal, Bangladesh and Pakistan. In: Climate change and disaster risk management. Springer, Berlin, Heidelberg, pp 149–167
8. Bustos P (2011) Trade liberalization, exports, and technology upgrading: evidence on the impact of MERCOSUR on Argentinian firms. Am Econ Rev 101(1):304–340
9. Charfeddine L (2017) The impact of energy consumption and economic development on ecological footprint and CO_2 emissions: evidence from a Markov switching equilibrium correction model. Energy Econ 65:355–374
10. Charfeddine L, Mrabet Z (2017) The impact of economic development and social-political factors on ecological footprint: a panel data analysis for 15 MENA countries. Renew Sustain Energy Rev 76:138–154
11. Chaudhry QUZ (2017) Climate change profile of Pakistan. Asian Development Bank
12. Cole MA (2004) Trade, the pollution haven hypothesis and the environmental Kuznets curve: examining the linkages. Ecol Econ 48(1):71–81
13. Copeland BR (2013) Trade and the environment. In: Palgrave handbook of international trade. Palgrave Macmillan, London, pp 423–496
14. Copeland C (2010) Air quality issues and animal agriculture: a primer. Anim Agric Res Prog 1
15. Destek MA, Ulucak R, Dogan E (2018) Analyzing the environmental Kuznets curve for the EU countries: the role of ecological footprint. Environ Sci Pollut Res 25(29):29387–29396
16. Dickey DA, Fuller WA (1979) Distribution of the estimators for autoregressive time series with a unit root. J Am Stat Assoc 74(366a):427–431
17. Eckstein D, Künzel V, Schäfer L, Winges M (2019) Global climate risk index 2020. Germanwatch Available at: https://germanwatch.org/sites/germanwatch.org/files/20-2-01e%20Global,20
18. Eskeland GS, Harrison AE (2003) Moving to greener pastures? Multinationals and the pollution haven hypothesis. J Dev Econ 70(1):1–23
19. Galli A, Kitzes J, Niccolucci V, Wackernagel M, Wada Y, Marchettini N (2012) Assessing the global environmental consequences of economic growth through the ecological footprint: a focus on China and India. Ecol Ind 17:99–107
20. Grossman GM, Krueger AB (1991) Environmental impacts of a North American free trade agreement (No. w3914). National Bureau of economic research
21. Kasman A, Duman YS (2015) CO2 emissions, economic growth, energy consumption, trade and urbanization in new EU member and candidate countries: a panel data analysis. Econ Model 44:97–103
22. Katircioglu S, Katircioğlu S, Altinay M (2018) Interactions between tourism and financial sector development: evidence from Turkey. Serv Ind J 38(9–10):519–542
23. Lee JW (2013) The contribution of foreign direct investment to clean energy use, carbon emissions and economic growth. Energy Policy 55:483–489
24. Mrabet Z, Alsamara M (2017) Testing the Kuznets Curve hypothesis for Qatar: a comparison between carbon dioxide and ecological footprint. Renew Sustain Energy Rev 70:1366–1375
25. Nabi I (2012) Pakistan's quest for a new growth vent: lessons from history. Lahore J Econ 17:243
26. Network GF (2019) Global footprint network. Obtenido de Global Footprint Network: http://www.footprintnetwork.org.online.Accessed, pp 1–10
27. Omri A, Mabrouk NB, Sassi-Tmar A (2015) Modeling the causal linkages between nuclear energy, renewable energy and economic growth in developed and developing countries. Renew Sustain Energy Rev 42:1012–1022

28. Ozturk I, Al-Mulali U, Saboori B (2016) Investigating the environmental Kuznets curve hypothesis: the role of tourism and ecological footprint. Environ Sci Pollut Res 23(2):1916–1928
29. Panayotou T (1997) Demystifying the environmental Kuznets curve: turning a black box into a policy tool. Environ Dev Econ 465–484
30. Perron P (1990) Testing for a unit root in a time series with a changing mean. J Bus Econ Stat 8(2):153–162
31. Pesaran MH, Shin Y, Smith RJ (2001) Bounds testing approaches to the analysis of level relationships. J Appl Econometr 16(3):289–326
32. Petroleum B (2019) BP statistical review of world energy report. BP, London, UK
33. Rudolph A, Figge L (2017) Determinants of ecological footprints: What is the role of globalization? Ecol Ind 81:348–361
34. Shahbaz M, Nasreen S, Abbas F, Anis O (2015) Does foreign direct investment impede environmental quality in high-, middle-, and low-income countries? Energy Econ 51:275–287
35. Solarin SA, Al-Mulali U, Musah I, Ozturk I (2017) Investigating the pollution haven hypothesis in Ghana: an empirical investigation. Energy 124:706–719
36. Udemba EN (2020) A sustainable study of economic growth and development amidst ecological footprint: new insight from Nigerian Perspective. Sci Total Environ 732:139270
37. Udemba EN (2020) Ecological implication of offshored economic activities in Turkey: foreign direct investment perspective. Environ Sci Pollut Res 27(30):38015–38028
38. Udemba EN, Güngör H, Bekun FV (2019) Environmental implication of offshore economic activities in Indonesia: a dual analyses of cointegration and causality. Environ Sci Pollut Res 26(31):32460–32475
39. Udemba EN (2020) Mediation of foreign direct investment and agriculture towards ecological footprint: a shift from single perspective to a more inclusive perspective for India. Environ Sci Pollut Res 1–18
40. Udemba EN, Agha CO (2020) Abatement of pollutant emissions in Nigeria: a task before multinational corporations. Environ Sci Pollut Res 1–11
41. WDI T (2019) World development indicators (DataBank)
42. Wang Y, Kang L, Wu X, Xiao Y (2013) Estimating the environmental Kuznets curve for ecological footprint at the global level: a spatial econometric approach. Ecol Ind 34:15–21
43. World Bank Group (2018) Doing Business 2018: reforming to create jobs: comparing business regulation for domestic firms in 190 economies. Internation Bank for Reconstrucion and Development/World Bank
44. Zarsky L (1999) Havens, halos and spaghetti: untangling the relationship between FDI and the environment. Foreign Direct Invest Environ 47–73
45. Zhan J (2020) Covid-19 and investment—an UNCTAD research round-up of the international pandemic's effect on FDI flows and policy. Trans Corporations J 27(1)

Printed in the United States
by Baker & Taylor Publisher Services